Elements Cycling and Plants' Physiological Characteristics: A Soil–Plant–Atmosphere Continuum Perspective

Elements Cycling and Plants' Physiological Characteristics: A Soil–Plant–Atmosphere Continuum Perspective

Guest Editors

Ling Zhang
Zhi Li

MDPI

Basel • Beijing • Wuhan • Barcelona • Belgrade • Novi Sad • Cluj • Manchester

Guest Editors
Ling Zhang
College of Forestry
Jiangxi Agricultural
University
Nanchang
China

Zhi Li
College of Forestry
Henan Agricultural
University
Zhengzhou
China

Editorial Office
MDPI AG
Grosspeteranlage 5
4052 Basel, Switzerland

This is a reprint of the Special Issue, published open access by the journal *Life* (ISSN 2075-1729), freely accessible at: https://www.mdpi.com/journal/life/special_issues/TAT9V093H3.

For citation purposes, cite each article independently as indicated on the article page online and as indicated below:

Lastname, A.A.; Lastname, B.B. Article Title. *Journal Name* **Year**, *Volume Number*, Page Range.

ISBN 978-3-7258-5627-5 (Hbk)
ISBN 978-3-7258-5628-2 (PDF)
https://doi.org/10.3390/books978-3-7258-5628-2

Cover image courtesy of Ling Zhang

Contents

About the Editors

Ling Zhang

Ling Zhang is affiliated with the College of Forestry, Jiangxi Agricultural University. His research interests are centered on sustainable forest management, forest ecology, and global change biology. He is actively involved in projects that integrate field survey data to improve the accuracy of forest resource inventories and monitoring. His work contributes to a deeper understanding of forest ecosystem dynamics and biodiversity conservation, and he has published several research articles in these fields, advancing the practical application of scientific knowledge in forestry.

Zhi Li

Zhi Li is affiliated with the College of Forestry, Henan Agricultural University. His research focuses on forest cultivation, silviculture, and the physiological ecology of trees. He is engaged in projects aimed at enhancing tree species' growth, stress resistance, and adaptive management strategies in the context of environmental change. Through his research, he has made significant contributions to the development of sustainable silvicultural practices and has successfully disseminated his findings through various scientific publications, supporting the advancement of forestry science and its application.

Preface

This Reprint, entitled "Elements Cycling and Plants' Physiological Characteristics: A Soil–Plant–Atmosphere Continuum Perspective", brings together a collection of research and review articles that explore the intricate dynamics of elemental cycling and the resultant physiological responses of plants within the integrated soil–plant–atmosphere system. The subject encompasses a wide scope, from molecular and genetic adaptations in various plant species, including trees, crops, and microalgae, to large-scale ecosystem assessments of phytomass and carbon stocks. The primary aim of this Reprint is to enhance our understanding of how plants perceive, respond to, and modify their environment through physiological and biogeochemical processes. The motivation for compiling this work stems from the urgent need to address global challenges such as nutrient deficiency, climate change, and sustainable agricultural and forestry practices. By integrating studies on rootstock-mediated nutrient tolerance, genetic diversity, microbial interactions, soil community shifts, and innovative waste upcycling, this volume seeks to provide a holistic view of plant–environment interactions. It is addressed to researchers, academics, and advanced students in plant physiology, ecology, agricultural sciences, and environmental biology, offering them a valuable resource of contemporary findings and methodologies that bridge fundamental research with practical applications for ecosystem management and agricultural sustainability.

Ling Zhang and Zhi Li
Guest Editors

life

MDPI

Article

Assessment of Phytomass and Carbon Stock in the Ecosystems of the Central Forest Steppe of the East European Plain: Integrated Approach of Terrestrial Environmental Monitoring and Remote Sensing with Unmanned Aerial Vehicles

Vasiliy Slavskiy [1,*], Sergey Matveev [1], Sergey Sheshnitsan [1], Daria Litovchenko [1], Maxim Viktorovich Larionov [2,*], Anton Shokurov [3,4], Pavel Litovchenko [1] and Nikolay Durmanov [5]

1 Faculty of Forestry, Voronezh State University of Forestry and Technologies named after G.F. Morozov, 8 Timiryazev Street, 394087 Voronezh, Russia; lisovod@bk.ru (S.M.); sheshnitsan@gmail.com (S.S.); timashchuk90@mail.ru (D.L.); pawellpv@yandex.ru (P.L.)

2 Department of Bioecology and Biological Safety, Institute of Veterinary Medicine, Veterinary and Sanitary Expertise and Agricultural Safety, Federal State Budgetary Educational Institution of Higher Education Russian Biotechnological University (ROSBIOTEC'H University), 1 Volokolamsk Highway, 125080 Moscow, Russia

3 Computational Methods Laboratory, Mechanics and Mathematics Faculty, Lomonosov Moscow State University, Leninskiye Gory 1, Main Building, GSP-1, 119991 Moscow, Russia; shokurov.anton.v@yandex.ru

4 Interdisciplinary Scientific and Educational School of Moscow University "Brain, Cognitive Systems, Artificial Intelligence", Leninskiye Gory 1, Main Building, GSP-1, 119991 Moscow, Russia

5 Ministry of Science and Higher Education of the Russian Federation, 11 Tverskaya Street, 21 Bryusov Lane, 125009 Moscow, Russia; nikolaydurmanov@gmail.com

* Correspondence: slavskiyva@yandex.ru (V.S.); m.larionow2014@yandex.ru (M.V.L.)

Abstract: The rapid and accurate estimation of aboveground forest phytomass remains a challenging research task. In general, methods for estimating phytomass fall mainly into the category of field measurements performed by ground-based methods, but approaches based on remote sensing and ecological modelling have been increasingly applied. The aim is to develop the scientific and methodological framework for the remote sensing estimation of qualitative and quantitative characteristics of forest stands, using the combination of surveys and machine learning models to determine phytomass of forest stands and calculate the carbon balance. Even-aged stands of different tree species growing in the forest steppe zone of the East European Plain were chosen as test objects. We have applied the modernized methodological approaches to compare and integrate forest and tree stand characteristics obtained by ground-based and UAV-based comprehensive surveys; additionally, we developed computer vision models and methods for determining the same characteristics by remote sensing methods. The key advantage of the proposed methodology for remote monitoring and carbon balance control over existing analogues is the minimization of the amount of groundwork and, consequently, the reduction inlabor costs without loss of information quality. Reliable data on phytomass volumes will allow for operational control of the forest carbon storage, which is essential for decision-making processes. This is important for the environmental monitoring of forests and green spaces of various economic categories. The proposed methodology is necessary for the monitoring and control of ecological–climatic and anthropogenic–technogenic transformations in various landscapes. The development is useful for organizing the management of ecosystems, environmental protection, and managing the recreational and economic resources of landscapes with natural forests and forest plantations.

Keywords: forest phytocoenoses; forest landscapes; canopy layer; morphometric parameters of trees; forest survey; phytomass; carbon stock; terrestrial environmental monitoring; remote sensing; vegetation indices; state of vegetation; landscape bioindication; synergy of methodological methods and approaches; machine/deep learning

1. Introduction

Global climate change has recently emerged as a serious long-term economic and environmental issue of universal concern [1,2]. Forest ecosystems play a significant role in the global carbon cycle; this has been confirmed by numerous scientific studies in recent decades [3–6].L. Chen et al. [7], C.I. Braga et al. [8], Y. Zhou et al. [9], A. Zielonka et al. [10], and other authors concluded that forest ecosystems contain approximately 80% of terrestrial carbon. It has been established that forest ecosystems play a significant role in reducing the concentration of greenhouse gases and mitigating global climate change [11–17].

Taking into account intensive forest management [18–20], deterioration of meteorological–climatic [21–24], and ecological [25–29] environmental parameters, the assessment of the bioproductivity of tree stands and forests [30–34] seems to be an urgent scientific and practical task. The long-term accumulation of stored carbon in forest ecosystems results in a biospheric effect related to the absorption of greenhouse gases, counteracting changes in the climate system of the planet [35,36]. Estimates of forest carbon stocks typically identify five primary pools: aboveground phytomass, belowground phytomass, deadwood (including standing and fallen dead wood), litter, and soil organic matter [3]. Aboveground phytomass is the key indicator of ecosystem productivity, forming qualitative and quantitative indicators of forest stands [37], as well as providingthe basis for the release of oxygen and carbon sinks [38]. Furthermore, determining the amount of carbon in soil and forest vegetation below ground is extremely difficult (and often impossible) when calculating plantation carbon stocks using remote sensing. The problem also lies in the obvious shortcomings of existing methods for the environmental monitoring of phytocoenoses through remote sensing.

The monitoring of the emissionand absorption of greenhouse gases by means of ground-based and remote sensing methods is extremely relevant and widely used in scientific and applied research [35]. Knowledge in the field of forest carbon accounting is currently increasing. However, different methodological approaches [38–43] and computational models [44–48] have been found to produce inconsistent results, highlighting the need for further methodological improvements.

This requires obtaining reliable information on the quantitative characteristics of forest stands and calculating the amounts of live and dead phytomass. Reliable estimation of phytomass in the context of climate change can provide a theoretical basis for the study of carbon cycling in terrestrial ecosystems [17,49], which has a crucial role in the response of forest ecosystems to greenhouse gas emissions [16,50]. The majority of authors [51–53] believe that a reliable assessment of aboveground forest phytomass will enhance the efficiency of management decisions in forest protection, conservation, and regeneration.

The most accurate method for estimating phytomass is in situ inventory with ground-based forest inventory methods [54]. The main disadvantage of the ground-based monitoring is high labor intensity, as this type of work involves a significant number of field surveys being conducted over an entire study area [55]. In addition, the problems with ground-based monitoring are related to the remoteness of the number of study sites from roads and settlements, the difficulty of moving across the study area, and often the inability to reach the forest area. This is especially true for the Central Forest Steppe zone of the East European Plain.

Rapid and accurate estimation of the aboveground forest phytomass remains a challenging research task [56,57]. In general, methods for estimating phytomass fall mainly into the category of field measurements made by ground-based methods, but remote sensing [58,59] and ecological modelling approaches have been increasingly applied [60,61].

It should be noted that remote sensing methods are of primary importance for over 70% of the Russian Federation's territory [62]. This is the optimal (and sometimes the only) method of obtaining statistically reliable and current information about forests for geographically remote and hard-to-reach areas [55].

Currently, satellite monitoring and assessment methods of forest states and structures are widespread in Russia and are used to estimate carbon balance [63,64]. With proper attribution of natural (disturbance) and anthropogenic (degradation) losses [65], the research has the potential to inform decision-making by governments across. It is well known that management and organizational decisions in relation to the rationalization of forest management, afforestation, the creation of plantings, and the environmental and economic control of plantings of different target categories are not sufficiently substantiated. This is observed in different administrative territories and geographical regions of the world. Focusing on research material on this issue is characterized by relevance and has practical benefits. Optical remote sensing images of Earth at various spatial, spectral, and temporal resolutions are freely available and are used to assess forest stands at various scales. This is reflected in the scientific works of many researchers [66–69]. Despite the obvious advantages of obtaining spatial data from satellite systems, this method of assessing forest stand phytomass has a significant disadvantage: the determination of forest tree stand characteristics is performed by analyzing the canopy of stands, but the resolution of satellite images does not provide detailed information. In addition, deadwood and understory vegetation are not taken into account. Assessments of carbon sequestration in the plantations often does not take into account the layering and structure of the forest stand. Carbon stock calculations are often very fragmented and based on approximate calculations.

Therefore, it is necessary for spatial data acquisition sources (camera type used, spatial resolution, survey age, etc.) to be enhanced. New methodologies and technologies need to be developed and implemented for the objective accounting of the carbon sink, allowing for the rapid estimation of phytomass in the forest stands, including the use of artificial intelligence (AI) [56]. Lidar technology based on the collection and analysis of data obtained using the laser scanning is widely used in the forestry sector [70]. Particularly, it can be successfully applied to improve our understanding of the three-dimensional structure of the forest environment [71]. It should be noted that there is practically lack of data on carbon sequestration in the lower layers of the forest phytocoenoses (including undergrowth and undergrowth) of the Central Forest Steppe of the East European Plain. In this regard, modern data are required on the state and productivity of forest ecosystems with differentiation according to different components of phytocoenoses.

Many researchers and scientists [72–77] have noted significant prospects for the automatic decoding of forest stands and analysis of forest ecosystems, including by laser scanning methods [78,79]. Specific vertical structuresare successfully recognized by the pattern of obtained dense point cloud and its shape using Lidars, including those based on UAVs [80]. In our study we used a combination of Lidar, RGB, and hyperspectral cameras. The effectiveness of the joint use of various types of sensors has been confirmed by many authors [81–84]. Their use corresponded to the purpose, theme, and problems of the work under consideration, as well as the idea of ensuring the synergy of advanced methodological approaches and technical means. This was necessary in accordance with the scientific idea and the specifics of our work.

Thus, the application of integrated ultra-high-resolution equipment during aerial survey [85,86] significantly increases informative and expands functional capabilities in phytomass assessment; it allows all silvicultural, bioecological, and tree stand characteristics to be identified, enabling us to determine carbon sinks in forests. Remote sensing is an extremely relevant and rapidly developing research area in the forestry industry. Remote sensing methods are widely in demand for use in assessing and monitoring forest stands.

The aim of the work is to develop the scientific and methodological basis for remote estimation of quantitative and qualitative characteristics of tree stands and ecosystems using colored light (RGB), Lidar, and hyperspectral surveys based on UAV and the presently widely accessible machine learning models (artificial intelligence)to determine aboveground phytomass of the forest stands and phytocoenoses.

Our proposed methodological approach for the remote ecological monitoring of the productivity of aboveground phytomass of forest ecosystems based on UAVs will provide more accurate and objective information about the state and structure of phytocoenoses. This methodological approach makes it possible to determine the carbon stock in natural forest stands and plantings and to carry out an accounting of forest resources. Our approach also allows us to monitor fire, environmental, and sanitary safety in forests.

2. Materials and Methods

2.1. Study Area

Considering the diversity of the forest vegetation in the Central Forest Steppe, this study assessed forest stands growing in the forest steppe zone using the example of the Suburban Forestry of the East European Plain (using the example of the Voronezh Region). The objects of the research were selected based on similar phytocoenotic, ecological, and physical–geographical conditions of the Central Forest Stepperegion. This ensured that the sample was representative and that reliable results were obtained for the general population. The location of the study sites is shown in Figure 1.

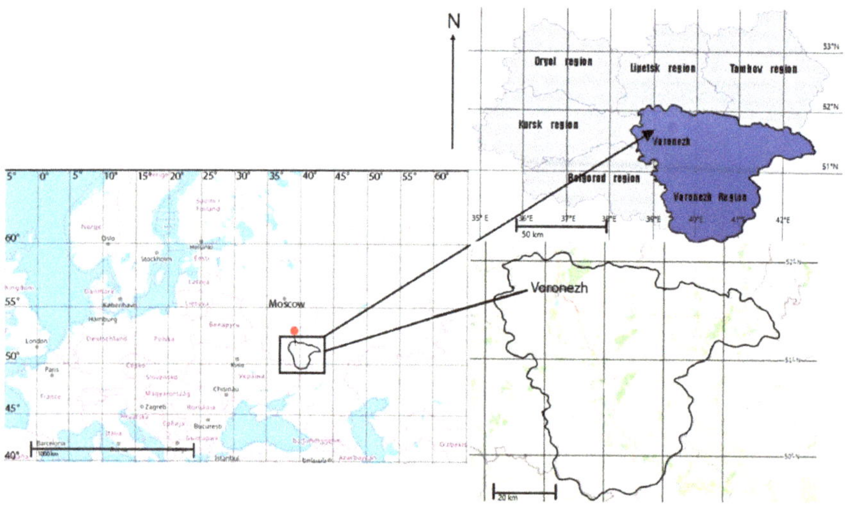

Figure 1. Scheme of the study area location.

The research program included the following steps:

1. Establishing sampling plots (Figure 2) in even-aged stands of different tree species composition and conducting the complete inventory on them with measurement of tree coordinates, species identification, trunk height and diameter, crown diameter, and tree state using ground-based inventory methods.
2. Performing aerial surveys using the RGB, hyperspectral, and Lidar imagery from the UAVs to determine similar silvicultural and taxation characteristics using remote sensing methods.
3. Generating training samples, designing computer vision models, and developing methods for determining the forest tree stand characteristics to calculate the phytomass of the forest stands using remote sensing methods.
4. Comparing and integrating forest and taxation parameters of the stands obtained during ground-based inventories and parameters of the forest areas obtained using the RGB, hyperspectral, and UAV-based Lidar surveys.
5. Application of the mathematical statistics and modelling methods to assess the accuracy of the results obtained, to verify the model validity, and to verify the developed algorithms.
6. Development of the methodology for remote sensing of aboveground phytomass of forest stands for subsequent calculation of carbon stock in the forest area.

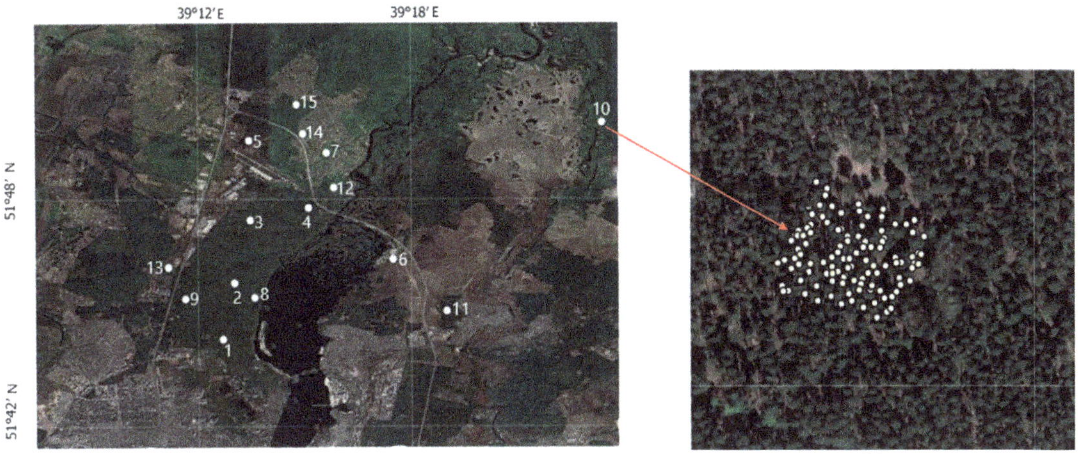

(a) (b)

Figure 2. Location of trial plots on the territory of suburban forestry (numbers indicate the numberssampling plots): (**a**) sampling of the sites location; (**b**) placement of trees on SP 10.

Fifteen sample plots (50 × 50 m) were established in the coniferous, deciduous, and mixed forest stands located in different forest growth conditions with moderately moist (fresh) types according to P.S. Pogrebnyak and in the different forest types according to V.N. Sukachev classification [87]. We studied even-aged (conditionally even-aged) stands growing in the following forest growth conditions: A_2 (fresh pine forest), B_2 (fresh pine–oak forest), C_2 (fresh pine forest with birch, aspen, and oak participation), C_2D (fresh pine forest with oak participation), D_2 (fresh oak forest). The age of the forest stands was 80–100 years. The tree stand characteristics of the studied stands are shown in Table 1.

Table 1. Species composition, tree stand characteristics, and forest growth conditions of the sample plots.

No of Sample Plot	Location	Composition of Tree Species, %	Age, Years	DBH, cm	H, m	f	Volume Stock, m^3 ha^{-1}	FGC
1	Quarter 44; site 26	70% Scots pine 30% English oak	90	35	27	0.6	266	C_2
2	Quarter 51; site 44	80% Scots pine 20% English oak	90	31	28	0.6	220	C_2
3	Quarter 27; site 14	90% English oak 10% Linden	90	30	26	0.6	250	C_2D
4	Quarter 11; site 25	80% English oak 20% English oak	90	28	24	0.6	210	D_2
5	Quarter 6; site 11	100% European birch	85	40	28	0.6	210	C_2D
6	Quarter 110; site 13	90% European birch 10% Scots pine	75	22	23	0.7	190	A_2
7	Quarter 9; site 28	80% Scots pine 20% European birch	90	30	26	0.7	310	C_2
8	Quarter 46; site 10	80% Scots pine 20% English oak	90	32	28	0.8	350	B_2
9	Quarter 48; site 44	80% Scots pine 20% Scots pine	90	30	24	0.8	330	A_2
10	Quarter 6; site 3	100% Scots pine	90	28	25	0.7	320	A_2

Table 1. *Cont.*

No of Sample Plot	Location	Composition of Tree Species, %	Age, Years	DBH, cm	H, m	f	Volume Stock, m^3 ha^{-1}	FGC
11	Quarter 76; site 21	100% Scots pine	90	30	26	0.6	290	A_2
12	Quarter 78; site 7	100% Scots pine	90	31	28	0.7	350	B_2
13	Quarter 48; site 79	100% Aspen	95	32	26	0.7	290	C_2D
14	Quarter 8; site 10	100% Aspen	100	30	26	0.6	230	C_2D
15	Quarter 48; site 17	100% Scots pine	95	32	28	0.7	350	B_2

Note: H—mean height; DBH—mean diameter on breast height; f—density of tree placement in the PRP (corresponds to the canopy density); FGC—forest growth conditions.

2.2. Ground-Based Forest Inventory

The full count was performed in all sample plots to determine tree species and the coordinates of individual trees, trunk diameter, tree height, crown diameter, and tree health state. Undergrowth, shrub layer, and the degree of coarse woody debris were identified and recorded.

The fieldwork was conducted utilizing modern geodetic and forest inventory equipment, including the SOUTH G2 GNSS receiver, Haglof Vertex Laser Geo altimeter. The methods employed were in accordance with the prevailing norms and legal frameworks in Russia [88–90].

Undergrowth and shrub layer were taken into account on the basis of the reforestation rules [91] and the forest management guidelines [92]. Undergrowth density degree (dense—over 8000 pcs. ha^{-1}, medium—2–8000 pcs. ha^{-1}, sparse—up to 2000 pcs. ha^{-1}) and size categories (1—up to 0.5 m; 2—0.6–1.5 m; 3—over 1.5 m) were determined and species determination was performed. The shrub layer was estimated according to the following criteria: density degree (dense—more than 5000 pcs. ha^{-1}, medium—2–5000 pcs. ha^{-1}, sparse—less than 2000 pcs. ha^{-1}), size (1—up to 1 m; 2—1–2 m; 3—more than 2 m), and species affiliation [90].

2.3. Estimation of the Aboveground Phytomass and Its Fractions

The estimation of aboveground phytomass stocks and its fractions (stem, branches, leaves) was conducted using allometric Equation (1) for each tree based on data on stem diameter at 1.3 m and tree height [93]:

$$\ln P_i = a_0 + a_1 \ln H + a_2 \ln DBH, \quad (1)$$

where P_i—phytomass stocks and its fractions (i; stem with bark, branches, needles/leaves), kg dry weight; H—tree height, m; DBH—stem diameter at 1.3 m, cm; a_0, a_1 and a_2—species-specific coefficients of allometric equation.

To calculate the phytomass of pear (*Pyrus communis* L.), apple (*Malus sylvestris* L.), and mountain (*Sorbus aucuparia* L.) ash, the equation for bird cherry (*Prunus padus* L.) was used as the most closely related systematic category. The aboveground phytomass of undergrowth and shrubs with the DBH less than 8 cm was calculated for each species using their height according to allometric Equation (2) [94]:

$$P_{ag} = a \times H^b, \quad (2)$$

where P_{ag}—aboveground phytomass, kg dry weight; H—tree height, m; a and b—species-specific coefficients of allometric equation.

Statistical analysis was performed using STATISTICA-13 [95]. Pearson correlation (3) and ANOVA analysis (4), as well as mathematical modelling techniques, were used [96].

The study employed regression analysis methods [97]. Parameter estimates for linear models, as well as those reduced to linear form, were calculated using least squares, and nonlinear least squares for all other cases. Student's t-test was used to test the null hypothesis of statistical significance for the estimated parameter values. Grouped data (by sample area) were analyzed using both linear and nonlinear mixed-effects models. The accuracy of linear mixed-effects models was assessed by conditional and marginal coefficients of determination (R^2) [98] (see Equation(5)). Pearson χ^2-test of agreement [99] was used to determine the consistency between the results obtained from the machine learning models and the actual distribution series obtained during fieldwork (see Equation(6)).

The following criteria were used to assess the quality of the models:

Correlation coefficient:

$$R = \frac{\Sigma\,(x_j - x)\cdot(y_j - y)}{\sqrt{\Sigma\,(x_j - x)\cdot(y_j - y)}^2} \tag{3}$$

Reliability of differences:

$$t = \frac{M_1 - M_2}{\sqrt{m_1^2 + \sqrt{m_2^2}}} \tag{4}$$

Coefficient of determination:

$$R^2 = 1 - \frac{\Sigma\,(y_j - y_j)2}{\Sigma\,(y_j - y_j)2} \tag{5}$$

Pearson agreement criterion:

$$\chi^2 = \sum \frac{(y_j - y_j)2}{y_j} \tag{6}$$

where M—mean value, m—standard error; x_i, y_i—actual value; $\hat{y}_i$—predicted value; $\hat{y}$—mean of actual values.

2.4. Flight and Survey Process

The data array was generated using the UAV payload (using combined equipment—RGB, Lidar, hyperspectral camera, and GNSS receiver). The Lidar and hyperspectral survey were performedfrom an unmanned aerial vehicle (UAV), Luftera LQ-5 (take-off weight 9.5 kg); digital photography in visible spectrum was performed from an unmanned aerial vehicle (UAV), Luftera LQ-4 (take-off weight 5.2 kg). The spatial resolution of the orthophotomaps was 2–2.5 cm/pixel.

The flights were performed at two altitudes, 60–70 m and 120–130 m, with horizontal speeds not exceeding 3 m/s and 6 m/s, respectively, to ensure sufficient overlap at the shooting interval of the 1 frame per second and to minimize blur. Hyperspectral imaging was performed using a Cubert S185 hyperspectral camera. Each frame taken by this camera is a combination of a panchromatic image in JPG format with a resolution of 1000 × 1000 pixels and a geometrically coinciding hyperspectral data cube of 50 × 50 = 2500 spectra. The spectral range is 450–950 nm; the number of spectral channels is 125.

2.5. Interpretation of the Tree Stand Characteristicsfrom High-Resolution Images

It should be noted that machine learning methods are not the primary focus of this paper. Moreover, the advances of machine learning methods have achieved such progress that they can be easily used in many practical problems. In this respect, the presented methods are well-known and universal, but require careful parameter tweaking for application for concrete forest plots.

For the neural network models, individual trees were labeled (tree species and its crown contour) on the small portion of each of the forest plots on the RGB geotiff-image. Data was prepared of the neural network classifier (bounding box rectangle were build)for each of the tree partitioning contours and data augmentation was applied for each of the images to improve robustness/generalizability. In particularsuch dataset was used to learn the model that was used to determine the tree species.

The neural network classifies the species of the individual trees. The condition and diameter of the crown were derived from this result. The computation of the hard-to-define taxation parameters was based on the identified relationships between the stand characteristics measured in the sample plots, the canopy digital model parameters, and the spectral characteristics of the aerial images. Training samples were generated from above mentioned labeled dataset and the YOLO neural network model was trained.

The characteristics of the forest stand were determined as follows:

- Tree canopy cover—by processing the RGB geotiffs;
- Tree height—by processing the Lidar imagery data;
- Precise tree location—by processing the Lidar imagery data;
- Tree species composition—from the RGB imagery using neural network models;
- Tree health state—with the hyperspectral data processing, based on the vegetation indices (NDVI, EVI, CVI);
- Crown diameter—using the RGB data processing and Lidar imagery;
- Stem diameter—by indirect indicators, based on the relationships identified using empirical relationships.

Traditional digital image processing techniques were used to separate trees from the background. The Lidar imagery uses geometric methods to interpolate the ground surface, namely by searching for the top of the forest canopy and then the height of the trees. Based on the Lidar imagery, the density of undergrowth and undergrowth in the stands was determined.

The complex of the calculation parameters includes the simplest indices NDVI, EVI, and CVI (Equations (7)–(9)), which allowed us to assess the condition of the tree stands (to identify the presence of dead or weakened trees).

$$NDVI = \frac{B_{\sim 0.8-0.9} - B_{\sim 0.63-0.75}}{B_{\sim 0.8-0.9} + B_{\sim 0.63-0.75}} \tag{7}$$

$$EVI = 2.5 \times \frac{(NIR - RED)}{NIR + 6RED - 7.5BLUE + 1} \tag{8}$$

$$CVI = \frac{NIR}{GREEN} \times \frac{RED}{CREEN} \tag{9}$$

where B—intensity of spectral radiation in the corresponding bands (μm); *NIR*—reflection in the near infrared region of the spectrum; *RED*—reflection in the red region of the spectrum; *BLUE*—reflection in the blue region of the spectrum; *GREEN*—reflection in the green region of the spectrum.

2.6. Remote Estimation of Aboveground Phytomass and Carbon Stocks

The main objective of the remote estimation methodology for aboveground phytomass and carbon stocks was to significantly minimize ground surveys. The methodology is based on a sequence of steps that include the implementation of automated methods for the remote estimation of vegetation using UAVs, based on the application of machine learning models (use of artificial intelligence).

1. UAV-based determination using machine learning models:
   - Study area's coordinates, terrain features, and delineation boundaries for forest inventory;
   - "Direct" silvicultural and stand characteristics, such as composition, height, crown diameter;
   - Condition and degree of debris, especially deadwood, in the stands.
2. Determination of "indirect" silvicultural stand characteristics such as age, stem diameter, canopy closure, and volume stock using reference materials and allometric equations, based on "direct" silvicultural and stand characteristics values.
3. Determination of the density of the undergrowth and shrub layer based on UAVs using the Lidar transect developed by the authors, as well as calculation of the phytomass of subordinate layers based on the generated tabular model.
4. Calculation of the carbon stock in the aboveground part of the forest stands (living and dead vegetation phytomass). Phytomass-to-carbon conversion factors were used according to IPCC guidelines [99]: 0.51 for coniferous tree species and 0.48 for broadleaved tree species.

3. Results

The methodological toolkit has been developed and a work algorithm has been defined that allows us to conduct remote assessment of the quantitative and qualitative forest characteristics with an accuracy that is comparable to that of traditional ground-based methods. The basis of the work algorithm is the integrated use of the Lidar, hyperspectral, and RGB survey materials using UAVs. This is necessary to obtain reliable information about the amount of aboveground phytomass and to calculate carbon reserves in the plantation by fraction. Information obtained online is extremely important for timely management decisions, for planning reforestation, and for environmental protection work.

3.1. Development of Technical and Methodological Tools for Determination of the Main Tree Stand Characteristics

3.1.1. State of Tree and Shrub Vegetation

Determination of tree health category is based on vegetation indices calculation. The vegetation indices are constructed from hyperspectral data according to well-known formulas, based on the entire spectrum. Figure 3 shows selected areas in the forest area for the contrasting comparison and mapping of the proposed methodology. Dead trees are marked in red (NDVI-0.051), while fully viable trees are marked in yellow (NDVI-0.694).

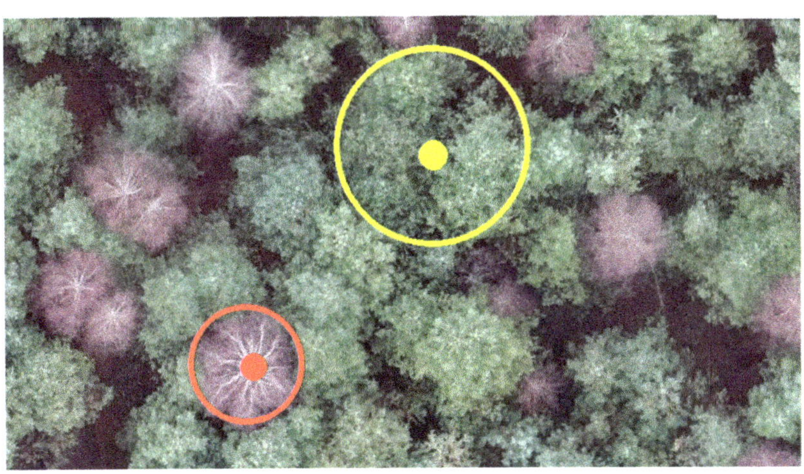

Figure 3. Selected areas for comparison of calculated indicators in the coniferous forest stands.

The point quantitative data of the vegetation indices reflecting the qualitative characteristic of the vegetation condition in the selected areas are collected in Table 2. Higher index values correspond to denser and greener vegetation in the analyzed fragment. The plots with dead trees have lower index values (NDVI less than 0.15). Vegetation indices are characterized by the greatest accumulation of chlorophyll in the plants. Therefore, the higher this value, the better the sanitary condition of the planting and the phytocenoses formed by them.

Table 2. Calculation of some indices for selected areas.

Vegetation Index	Indicator Value in the Red Area	Indicator Value in the Yellow Area
NDVI	0.0515	0.6944
CVI	1.1382	2.9996

Figure 4 shows the forest plot with trees of different health categories. The calculated NDVI layer is superimposed on the original image, whereby the presence of dead and viable trees in the area can be identified: yellow (tree stand, understorey, shrub layer) reflects viable vegetation, black reflects dense ground cover, and red reflects the reduced NDVI value and allows for the identification of dead trees.

(**a**) (**b**)

Figure 4. Fragment of images of trees with dead/lost foliage and healthy trees: (**a**) RGB photo; (**b**) superimposed NDVI calculation layer.

3.1.2. Determining Tree Coordinates and Measurement of Stem Height and Crown Diameter

The initial task of remote sensing data analysis (RGB complex, Lidar, and hyperspectral imagery) is to determine the location (coordinates) of trees in order to determine the main taxonomic characteristics of the tree stand.

The vegetation height matrices derived from the Lidar data are used for this purpose(Figure 5). The square window of fixed size is created in the center of each matrix cell; if the maximum of the window is reached at this point, the cell coordinates are declared to be tree coordinates, i.e., the tree coordinates coincide with the coordinates of their tops. Heights are then calculated at the derived points (Figure 6), and then the coordinates of the points are deduced from the coordinates of the las-file.

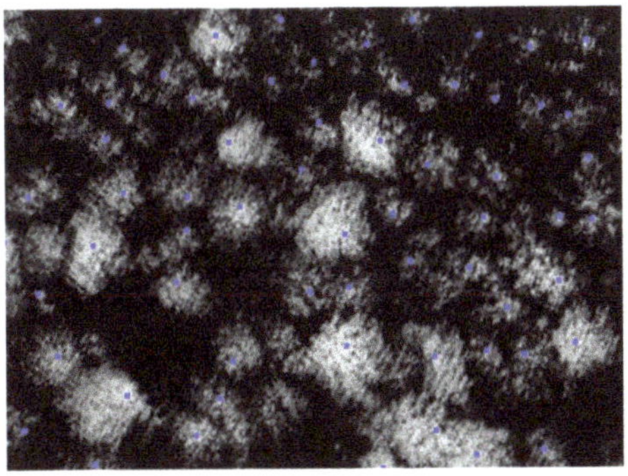

Figure 5. Maximums found on the elevation maps (marked in blue).

Figure 6. Formation of the height map (dots in green are trees, in red are their maxima).

Tree Species Identification

Tree species identification was performed based on the computer vision process, consisting of expert partitioning of the tree species composition (Figure 7), training samples, and generating machine learning models based on convolutional neural networks on the YOLO architecture. Training an object-detection model such as the YOLOv5 requires a dataset containing object images and the coordinates of the bounding boxes of the objects themselves.

The species composition of the forest stand was determined from RGB images using machine learning models. The algorithm for decoding species composition using artificial intelligence (based on neural networks) was conducted as follows:

1. Manual marking was completed—the operator carried out contour and analytical interpretation, as a result of which the species composition of the forest stand was determined.
2. Integration of manual marking data and RGB images.
3. Formation of training samples using neural network analysis.
4. Testing and calibration of machine learning models created based on neural networks.
5. Hyperspectral data were used to clarify the species identity and correct the breed composition.

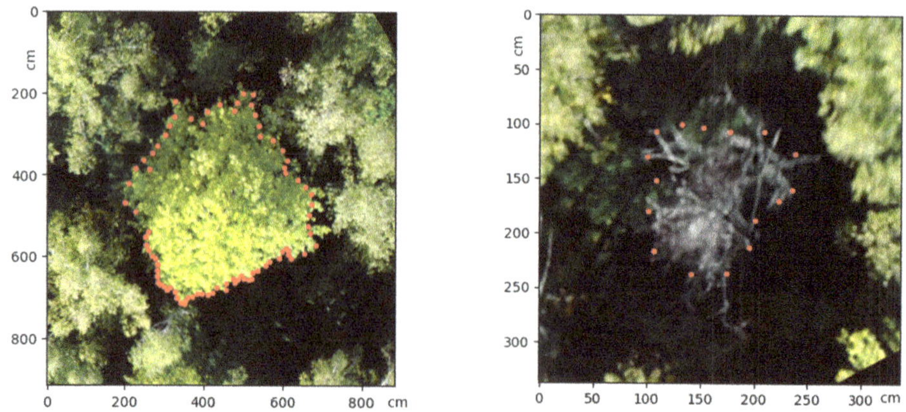

(**a**) (**b**)

Figure 7. Outline from expert mark-up: (**a**) live tree; (**b**) dead tree(verification of crown contours is indicated by red dots).

Determination of Crown Diameter

The tree crown diameter decoding process was based on the processing of point clouds obtained during the Lidar imagery processing. Knowing the coordinates (centers) of the trees, it is possible to determine both the heights and the crown diameters of individual trees. Pre-processing is performed manually in specialized point cloud software (e.g., QGIS, https://qgis.org/ru/site/ accessed on 8 May 2024) (Figure 8).

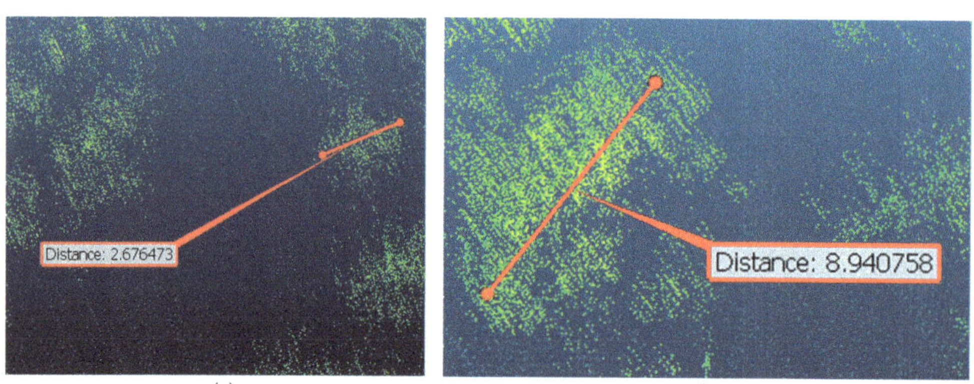

(**a**) (**b**)

Figure 8. Measurement of crown diameters (meters) from the Lidar data: (**a**) tree crown diameter on SP 13; (**b**) tree crown diameter on SP 3.

Following training of the AI-based on the neural network method, crown diameters were determined automatically.

Determination of Tree Height and Height of Crown Base

The process of determining tree heights was based on working with the Lidar point clouds. Each point cloud file contains information about the coordinates of the points, which allows us to determine the ground surface and the forest canopy; by subtracting one from the other, one candeterminethe stand height in meters (Figure 9).

Figure 9. Measurement of tree height (meters) from the Lidar data: (**a**) maximum tree height on SP 8; (**b**) maximum tree height on SP 10.

Automated tree height estimation involves calculating ground and vegetation surfaces, and then estimation of heights is performed by subtraction of the ground surface coordinate from the vegetation surface.

3.1.3. Shrub Layer and Understory Phytomass Assessment

The shrub layer and the understory in the stand are not always fully accounted for in carbon balance estimations [65,100–103]. However, the density and height of undergrowth can have a significant impact on the carbon content of phytomass. To improve the accuracy of determining the forest stand phytomass, the scale for accounting undergrowth and undergrowth by remote sensing methods using hyperspectral and the Lidar survey data was developed.

To calculate the amount of undergrowth using remote methods, a Lidar section was created (Figure 10), covering an area of 50 × 1 m. Based on the obtained value from Tables 3 and 4, the phytomass was calculated.

The scale was compiled on the basis of the experimental materials (field samples obtained both in this study and in earlier studies), with the application of mathematical modelling methods for interpolation and approximation of the results. The phytomass was calculated considering categories of density and height (see Tables 3 and 4).

The Lidar transect was performed using the developed tools with the web interface shown in Figure 10a. At least three Lidar transects should be made in the studied area (Figure 10b) and then the average number of shrubs and understory should be determined.

The scientific novelty of working with hyperspectral cubes is their processing algorithm. Clustering of spectra for each image is performed; here, 2500 spectra are grouped into several clusters with the k-means clustering algorithm. The resulting median spectra are compared with the spectra of reflected light from the plants, on the basis of which they are combined into clusters. The spectra belonging to the clusters that clearly do not correspond to the plant spectra are then excluded from further consideration, while the remaining spectra are involved in computer vison methods. In partitioning mode, training samples are generated from them. The example of the complex data processing is shown below (Figure 11).

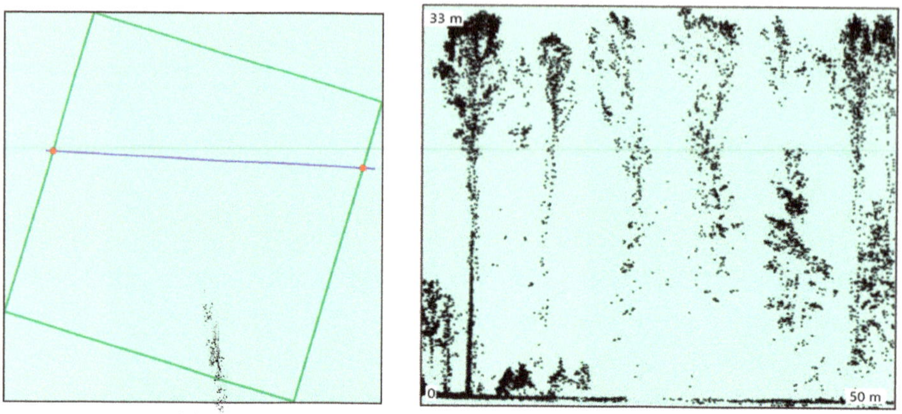

(**a**) (**b**)

Figure 10. Displaying of the Lidar data in the web interface: (**a**) general view and location of the Lidar section in the study area; (**b**) three-dimensional structure of the forest stand displayed as the Lidar transect.

Table 3. Reference values of the shrub phytomass in the stands growing in the fresh forest growth conditions, kg ha^{-1} dry weight (range of variation/mean value).

Criteria	Density Degree, pcs. ha^{-1}								
	Dense (More than 5000)			Medium (2–5000)			Sparce (Less than 2000)		
Number, pcs. transect^{-1}	>25			11–25			1–10		
Average height, m	>2.0	1.1–2.0	<1.0	>2.0	1.1–2.0	<1.0	>2.0	1.1–2.0	<1.0
Shrub phytomass in coniferous stands	600–4900 2600	125–1900 900	50–400 180	550–1800 1050	50–650 300	5–140 90	10–1100 400	5–250 100	1–50 20
Shrub phytomass in deciduous stands	1500–5200 3000	150–2650 1100	60–760 200	600–2500 1400	60–700 500	5–150 100	10–1500 650	5–550 200	1–50 25
Shrub phytomass in mixed stands	1250–5200 2950	150–2600 1050	50–600 190	600–1920 1260	50–650 400	5–150 100	10–1325 500	5–470 180	1–50 20

Table 4. Reference values of the understory phytomass in the stands growing in the fresh forest growth conditions, kg ha^{-1} dry weight (range of variation/mean value).

Criteria	Density Degree, pcs. ha^{-1}								
	Dense (More than 8000)			Medium (2–8000)			Sparce (Less than 2000)		
Number, pcs. transect^{-1}	> 40			11–39			1–10		
Average height, m	>1.5	0.6–1.5	<0.5	>1.5	0.6–1.5	<0.5	>1.5	0.6–1.5	<0.5
Understory phytomass in coniferous stands	5000–22,000 7500	500–9900 1900	100–900 450	1100–13,500 2500	90–6100 750	20–300 150	10–3950 550	5–1150 150	1–50 25
Understory phytomass in deciduous stands	3500–24,000 6000	380–5500 2000	100–800 400	800–18,000 2700	60–3500 700	20–250 130	10–5500 750	5–850 170	1–40 25
Understory phytomass in mixed stands	3200–16,000 5900	520–6200 1800	100–750 400	800–8400 2600	50–3200 700	30–250 130	10–3600 650	5–800 160	1–40 20

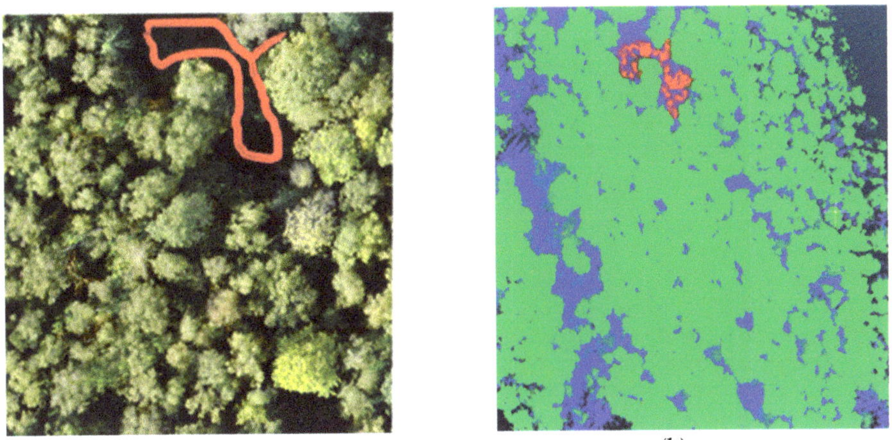

(**a**) (**b**)

Figure 11. Combining the RGB (**a**) and Lidar (**b**) images by common attributes.

The hyperspectral and Lidar devices provide separate categories of data to describe forests at individual tree level. The hyperspectral imagery contains meaningful plant reflectance characteristics or spectral features, while Lidar data enables the analysis of a canopy's structural properties. Combining these two data sources improves the quality of the forest mapping (Figure 12).

(**a**) (**b**)

Figure 12. Analyzed cross-section lines (**a**) and their mapping after processing the Lidar data (**b**) in stand studies.

This requires different data collection methods, appropriate processing algorithms, and classification and/or regression techniques. Remote sensing method integration aims to generate integrated information based on data with different spatial and spectral resolutions. These integrated data are more reliable and accurate than individual sources of information. Consequently, their integration improves the quality of the work and provides increased accuracy in calculating phytomass stock and estimating the carbon storage of the forest stands.

3.2. Determination of Correlations between Stand Characteristics in Order to Indirectly Identify the Deciphering Characteristics of the Forest Stand for Analytical Interpretation

Tree stem diameter was determined by indirect methods based on the empirical data and developed regression equations. Crown diameter and tree height have a diagnostic function and are in close correlation with the stem diameter at breast height. This makes it possible to develop regression models to estimate various characteristics of individual trees, which can then be used in aerial monitoring.

Preliminary correlation analysis of the data has shown that, for individual tree species, there is a fairly close relationship between stem diameter at 1.3 m height and crown diameter (R = 0.56 and 0.67 in coniferous and deciduous plantations, respectively) or tree height (R = 0.71 and 0.68); the strength of the relationship with the height parameter was generally higher than the corresponding relationship with crown diameter (Figures 13 and 14).

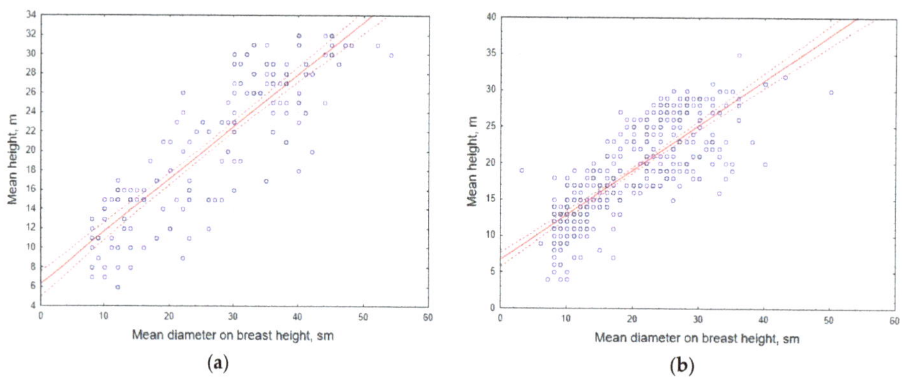

Figure 13. Correlation of stem diameter and tree height in coniferous (**a**) and deciduous (**b**) tree stands.

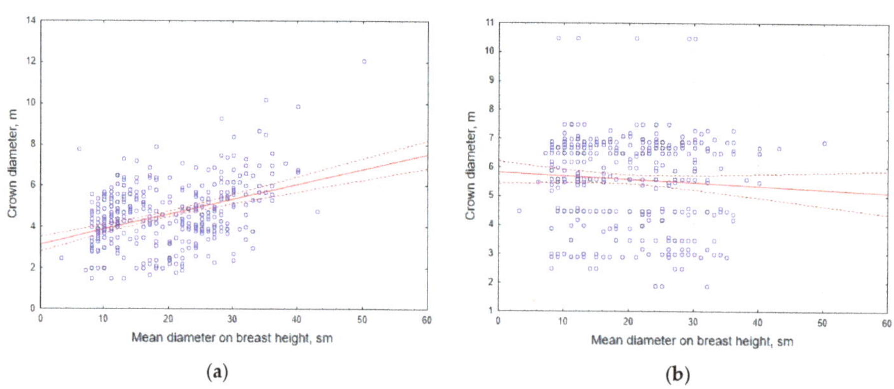

Figure 14. Correlation of crown diameter and stem diameter in coniferous (**a**) and deciduous (**b**) tree stands.

Tree height and crown diameter are essential for remote forest monitoring since they can be accurately determined from the RGB and Lidar imagery. Based on the results of the multivariate regression analysis, models were developed to determine tree stem diameter at 1.3 m using these most informative morphometric indicators of trees (see Equation(10)). The results of the calculation of the equation coefficients are summarized in Table 5.

$$Y = a_0 + a_1 H + a_2 D_{cr} \tag{10}$$

where Y—stem diameter at breast height 1.3 m (cm); H—tree height, m; D_{cr}—average crown diameter, m; a_0, a_1, and a_2—regression coefficients.

Table 5. Linear regression equation constants (a_0–a_2) for calculating stem diameter of different tree species.

Tree Species	Regression Coefficients			R	R^2	SE
	a_0	a_1	a_2			
Scots pine (*Pinus sylvestris* L.)	−2.7626 ± 0.1490	0.8739 ± 0.0123	2.2386 ± 0.0653	0.955	0.913	4.055
Oak (*Quercus robur* L.)	2.4860 ± 1.7332	1.0969 ± 0.0781	0.4070 ± 0.2163	0.634	0.402	9.039
Birch (*Betula pendula* L.)	−3.8368 ± 0.3637	0.6860 ± 0.0247	2.2120 ± 0.1015	0.920	0.846	4.223
Aspen (*Populus tremula* L.)	−3.6842 ± 0.4640	0.8152 ± 0.0626	1.9195 ± 0.2784	0.965	0.932	3.820

Species-specific coefficients are represented as a_n ± 95% confidence interval.

The calculated coefficient of determination (R^2) in the considered options exceeded 0.84 and confirmed the agreement of the linear regression models with the experimental data. The exception is oak ($R^2 < 0.5$), which requires an adjustment to the model used.

The results of the regression analysis improve the explanatory capacity of the models by 30–40% when several of the most significant morphometric indicators are used for prediction. Therefore, preference should be given to two- and multifactor models. Calculations based on the developed allometric models provide the acceptable error in determining stem diameter of almost all tree species (except for oak)—for aspen, pine, and birch, the equations provide the most accurate estimates of the considered taxation characteristics (Figure 15) and therefore could be applied in the analysis of the Lidar survey results.

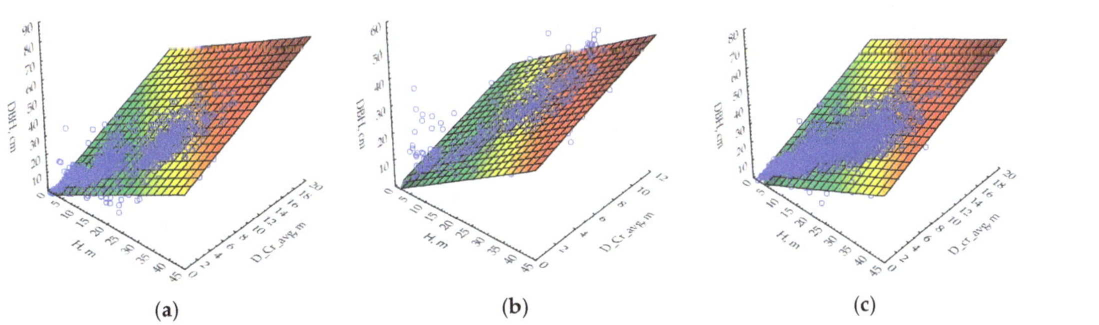

Figure 15. Relationship between stem diameter (DBH, cm) and mean crown diameter (D_cr_avg, m) and tree height (H, m) for main tree species in the Central Forest Steppe of the East European Plain for different tree species: (**a**) Scots pine (*Pinus sylvestris* L.); (**b**) birch (*Betula pendula* L.); (**c**) aspen (*Populus tremula* L.).

3.3. Validation of the Proposed Phytomass Estimation Toolkit

The next stage of the study comprised testing the toolkit developed to determine the phytomass of the forest stands. The comparison was made between the results obtained by interpretation of imagery and the results of the ground-based inventory of trees on the sample plots.

The estimation of the phytomass by remote sensing was carried out using data on species composition and stand health (based on the RGB and hyperspectral imagery), stem diameter (derived from the allometric equations), and stand height (derived from the Lidar imagery). Average data from Tables 4 and 5 were used to calculate understory and shrub phytomass. Additionally, dead wood was taken into account, based on the RGB and hyperspectral surveys.

The detailed data on the amount and fractional structure of phytomass obtained by ground-based inventory of sampling plots are shown in Table 6.

Table 6. Phytomass stocks of the forest stands (age of 90–100 years) on the sample plots (SP) in the fresh forest growth conditions of the Central Forest Steppe of the East European Plain (kg ha^{-1} dry weight).

No of SP	Tree Stand, Including							Shrub Layer	Understory	Total Phytomass
	Stems		Branches		Needles/Leaves		Total			
	M ± SE	Total	M ± SE	Total	M ± SE	Total				
1	494 ± 21	248,800	80 ± 6	40,368	15 ± 0.8	7342	296,510	418	6078	303,006
2	404 ± 23	163,127	73 ± 7	29,467	13 ± 0.9	5091	197,686	549	1954	200,189
3	513 ± 29	203,294	90 ± 6	35,634	7 ± 0.3	2644	241,572	461	3025	245,059
4	493 ± 29	206,855	92 ± 6	38,447	7 ± 0.4	2757	248,060	622	2121	250,804
5	644 ± 31	252,281	129 ± 8	50,576	14 ± 0.8	5467	308,324	168	926	309,418
6	224 ± 13	102,361	33 ± 2	15,031	7 ± 0.4	2974	120,366	142	75	120,583
7	345 ± 21	177,882	46 ± 4	23,482	14 ± 0.9	7309	208,672	954	1480	211,107
8	483 ± 28	274,535	86 ± 9	49,100	14 ± 0.8	7874	331,508	764	8044	340,317
9	288 ± 12	177,569	39 ± 3	23,832	12 ± 0.4	7400	208,801	562	998	210,361
10	288 ± 15	128,924	33 ± 2	15,003	12 ± 0.6	5526	149,452	210	324	149,987
11	315 ± 14	128,454	38 ± 2	15,388	14 ± 0.6	5584	149,426	510	3259	153,194
12	351 ± 15	180,927	37 ± 2	19,339	15 ± 0.7	7710	207,976	665	1018	209,659
13	378 ± 8	177,015	60 ± 2	28,023	10 ± 0.3	4781	209,819	347	142	210,308
14	362 ± 17	141,827	55 ± 4	21,487	9 ± 0.4	3352	166,667	281	246	167,193
15	414 ± 20	207,249	55 ± 4	27,474	18 ± 1.1	8774	243,497	381	2478	246,356

M—mean value of the sign; SE—standard error.

An analysis of Table 6 suggests that the maximum aboveground phytomass stocks in the forests of the Voronezh Region were in mixed oak–pine stands, reaching 303.0–340.3 t ha^{-1} (SP 1 and SP 8). Significant stocks of the aboveground phytomass (up to 309.4 t ha^{-1}) were also found in the birch forest with ash and linden. The lowest values of the phytomass were observed predominantly in the coniferous stands (dominated by Scots pine)—up to 153.2 t ha^{-1} (SP 10, SP 11).

Understory and shrub layer together contribute about 1–2% of total phytomass. The lowest contribution of the lower forest layers to the total forest stand phytomass was observed in mixed birch–pine forest stands—0.2% (SP 6, SP 13)—and the highest proportion of the total phytomass was in subcanopy layers of mixed oak–pine forest stands—2.6% (SP 8) (Figure 16).

Generally, the fraction in tall (over 2 m in height) undergrowth tends to be higher than in the comparable category of undergrowth. The undergrowth fraction varies over the wide range of two orders of magnitude, from 75 to 8044 kg ha^{-1}. Contribution of the undergrowth to the total phytomass on the investigated sample plots was not significant and did not exceed 0.5%. Its stocks vary from 142 kg ha^{-1} in mixed birch–pine forests to 954 kg ha^{-1} in mixed stands with Scots pine predominance.

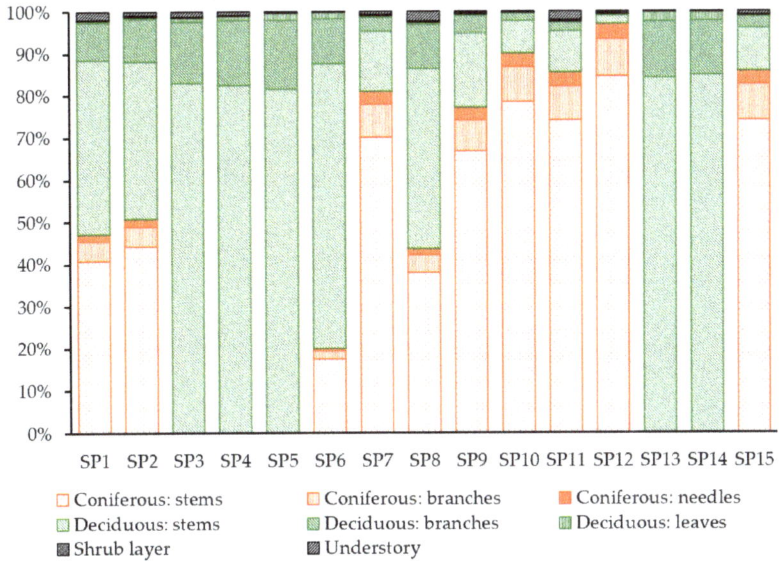

Figure 16. Fractional composition of the phytomass on the sample plots. Sample plot numbers are marked on the horizontal axis.

Stem wood accounts for at least 81% of the total phytomass of the stand. The analysis of the ratio of woody phytomass fractions clearly shows the pattern of increasing contribution of branches in mixed forests with the predominance of deciduous species, and deciduous forests—up to 15–16%—while the proportion of branches in coniferous forests accounts for less than 9–10% of total phytomass.

The statistical processing of the results was performed using the ANOVA (Table 7) to determine the significance of differences between the average values of the tree stand characteristics and the total amount of phytomass in stands, determined by ground-based and remote sensing methods (Figure 17).

Table 7. Significance of differences between the mean values of phytomass in the forest stands ($t_{a\text{-}b}$) determined by ground-based (a) and remote sensing (b) methods.

Type of Tree Stands	Sample Plots (Differences in Phytomass, a–b)															$t_{0.05}$
	1	2	3	4	5	6	7	8	9	10	11	12	13	14	15	
Coniferous									−1.6 *	−1.7 *	0.8 *	2.3			0.2 *	1.96
Deciduous			1.7 *	1.8 *	5.5								0.5 *	1.4 *		1.96
Mixed	5.2	−3.1				−2.6	1.8 *	0.6 *								1.96

* Insignificant differences between the means.

Significant differences (at the significance level of $p < 0.05$) between the average values of the total phytomass from the ground-based and remote sensing methods were found in 33% of the cases. However, the maximum difference (observed at SP 5) in the approximation of the data did not exceed 15%. In 66% of the cases, the average phytomass values determined by the different methods did not differ significantly from each other. This demonstrates the high accuracy of the information obtained by remote sensing, using the proposed technical–methodological approach.

Remote sensing performed on the basis of the proposed methodology is comparable in accuracy to ground-based methods for determining the carbon stock in aboveground phytomass. This is achieved by taking into account undergrowthand dead wood.

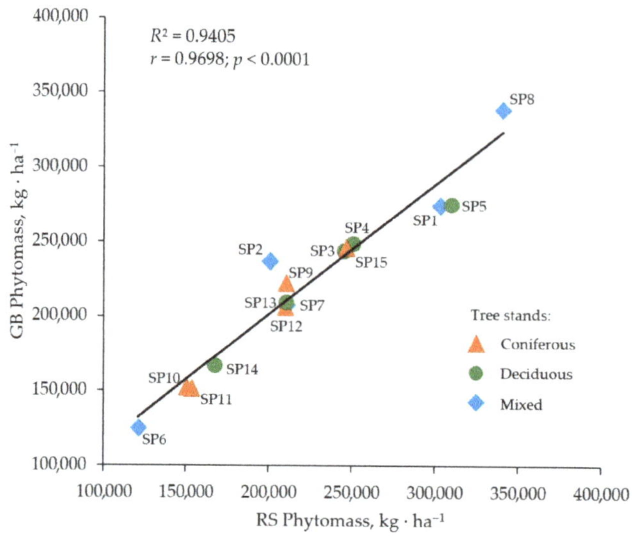

Figure 17. Total phytomass of different tree stand types on the sample plots (SP) determined by ground-based (GB) and remote sensing (RS) methods.

The similarity of phytomass values obtained by the ground-based and the remote sensing methods was statistically confirmed: Pearson criterion χ^2 (0.05; 5) = 2.79; $P(\chi^2) = 0.68 > 0.05$ was found to causeslight differences in the results of calculating phytomass by different methods. Consequently, we can reliably use the distance-based phytomass estimation technique to calculate the carbon storage of forest stands.

3.4. Calculation of Carbon Stock from Phytomass Measurements

The carbon stock calculations for the aboveground phytomass of the forest stands in the sample plots are summarized in Table 8.

Table 8. Aboveground carbon stock of the forest stands (age of 90–100 years) on the sample plots (SP) in the fresh forest growth conditions of the Central Forest Steppe of the East European Plain (kg ha^{-1}).

No of SP	Tree Stand, Including				Shrub Layer	Understory	Total Carbon Stock
	Stems	Branches	Needles/Leaves	Total			
1	123,128	19,794	3690	146,612	200	2918	149,730
2	80,952	14,427	2559	97,937	263	938	99,139
3	97,581	17,104	1269	115,954	221	1452	117,628
4	99,291	18,455	1323	119,069	299	1018	120,386
5	121,095	24,276	2624	147,995	81	444	148,520
6	49,766	7282	1457	58,504	68	36	58,608
7	89,823	11,767	3708	105,297	458	711	106,466
8	135,648	23,991	3947	163,586	367	3861	167,814
9	89,450	11,896	3748	105,093	270	483	105,846
10	65,411	7574	2808	75,793	101	156	76,049
11	65,061	7754	2836	75,652	245	1564	77,460
12	92,159	9843	3928	105,931	319	490	106,740
13	84,967	13,451	2295	100,713	166	68	100,948
14	68,077	10,314	1609	80,000	135	118	80,253
15	104,951	13,814	4460	123,225	183	1192	124,600

The total carbon stocks in the forest stands varied from 58.6 t ha^{-1} in birch-dominated stands (SP6), where minimal stem wood stocks were observed, to 167.8 t ha^{-1} in mixed pine-dominated stands (SP8). The shrub layer and understory contributed no more than 3.8 t C ha to the total carbon stocks. In most cases, the carbon stocks in their aboveground phytomass did not exceed 1.0 t ha^{-1}.

4. Discussion

The hyperspectral and Lidar data complement each other. The synergy between the technologies used can be referred to as 3D-imaging spectroscopy, which has been explored in the works of several authors [104,105]. The proposed methodology is primarily based on the use of the RGB and Lidar sensors estimating as much of the forest stand phytomass as possible and allows consideration of all carbon pools for the rapid and reliable calculation of the carbon storage [106–108].

Similar methods for remotely estimating stands and determining the accumulation of forest phytomass using the Lidar system were used for the tree species *Quercus semicarpifolia* (Sm.) and *Pinus roxburghii* (Sarg), when calculating the carbon stock in the subtropical and temperate forests of India [108–110]. The proposed method was based on the assessment of ground phytomass and carbon only for the tree layer, which is not appropriate and objective when assessing forest phytocoenosis as the whole. Please note that the machine learning models used are developed on a regional basis.

In monitoring the ecological state of vegetation and carbon reserves, great importance belongs to the qualitative and quantitative characteristics of the productivity of aboveground phytomass of individual plants and phytocoenoses. Moreover, in contrast to the traditionally implemented assessment of the carbon pool in the aboveground phytomass of tree stands [33,34,64], we have studied the parameters of the accumulation of aboveground phytomass and carbon in other tiers of the forest phytocoenoses. This also indicates the feasibility and objectivity of the assessment.

In addition to the required technical and software tools, adequate methodological apparatus is required. In our case, a comprehensive technical and methodological approach was used. We believe that the assessment required data from our ground-based monitoring on the morphometric characteristics of plants and their condition. This was confirmed by a correlation analysis. Positive correlations between the morphometric parameters of the woody plants were obtained. In particular, the parameters of tree height and crown diameter are important for remote environmental monitoring of the forest phytocoenoses. They can be determined by ground-based methods and remotely (from RGB images and through Lidar surveys). Multivariate regression analysis makes it possible to develop models for determining the diameter of the tree trunk, using morphometric parameters. The values of the species-specific coefficients for the number of the tree species are given. Priority should be given to two- and multifactor models when predicting the necessary morphometric parameters of ecologically and economically valuable plants. The validation of the proposed set of tools is disclosed and explanations are given for estimating phytomass using the example of the forest phytocoenoses.

The data obtained from ground-based monitoring and remote monitoring were compared differentially for the tree stands, undergrowth, shrub layer, and in general (Table 6). The major advantage of the methodology proposed for remote monitoring and control of the carbon storage compared to existing analogues is the minimization of the amount of groundwork and, consequently, the reduction inlabor costs without loss of information quality. Reliable information on the phytomass volumes will allow for operational control of forest carbon storage, which is essential for timely management decisionmaking. The proposed methodology is useful for analyzing the possibilities of sequestration of carbon and other elements in plant phytomass and soils in forested, agricultural, and populated areas. Of great importance is the application of the described methodology for studying the ecosystem-related, ecological–protective, and phytomeliorative properties of tree stands and other vegetation elements in relation to agrophytocoenoses.

Wu et al. [111], Reinmann et al. [112], Junfang et al. [113], and thenumber of other researchers [114–116] believe that the quantification of the aboveground forest phytomass and carbon cycle studies in general is a highly relevant and current issue, which will provide the theoretical basis for the study of the carbon cycle and global climate change. Obtaining information on the forest ecosystems and carbon storage will make it possible to identify the inverse relationship between the processes occurring in forests and climate change. It also seems promising to analyze the state of tree plantations and forests, the accumulation of their phytomass to consider the balance of the carbon and nutrient cycles, to assess degradation processes due to environmental and climatic transformations in the environment and in the forest natural ecosystems and cultural ecosystems [117–119]. Data on the formation of the phytomass of the forests and plantations can be used as the basis for environmental and climate–economic management of natural and cultural ecosystems. This is consistent with the results of other authors [120–123]. This seems possible in geographically, ecologically, and economically diverse areas. The presented methodology for integrated carbon monitoring allows us to identify the ecological state, the limits of sustainability, and the resource base of the corresponding natural phytocoenoses and agrophytocoenoses. The proposed universal methodology can be of decisive importance in the selection of plant species and varieties for afforestation, landscaping, horticulture, protection, and restoration of soil ecosystems, and the protection of and increase in the resource attractiveness of landscapes on the East European Plain and other territories.

5. Conclusions

The use of UAVs for the remote sensing of forest ecosystems in the Eastern European Plain is the issue that is becoming more and more important. Estimating the amount of carbon stored in forests is the critical undertaking. At present, there is no single system that integrates the collection, storage, and analysis of data on forest ecosystem health and structure, leading to significant uncertainties in current biomass assessment methods. It is crucial to enhance existing methods and develop new approaches for estimating forest phytomass and calculating carbon stocks.

The estimation of the carbon storage of the forests is based on the calculation of the phytomass of the stand and other components of the forest ecosystem. Meanwhile, woody phytomassis based on the application of allometric equations or conversion factors, defined as the ratio of total stand phytomass (including aboveground, belowground, and understory) to the stemwood stock. The conversion factors are differentiated according to the age of the stand and other taxonomic indicators obtained by ground-based forest inventory.

In the study area, on average, up to 81% of the total volume of formed phytomass is stem wood. To complete the assessment, the contribution of other layers of forest vegetation was also established. The undergrowth and shrub layer are responsible for the formation of aboveground phytomass in the volume of up to 2% of the total volume of aboveground phytomass of forest phytocenoses. Phytocenoses of the oak, oak–pine, pine, and birch forests are natural ecological frameworks for the ecosystems they form and for the ecosystems adjacent to them. The total carbon reserves in the aboveground phytomass of phytocenoses varied, depending on their composition. The total carbon reserve reached its maximum value (163.6 t ha^{-1}) in plantations of mixed species composition with the predominance of the pine.

To integrate and approximate data obtained during ground surveys and data obtained during remote monitoring based on the UAVs, technical and methodological tools have been developed. This toolkit allows you to analyze and decipher information. The technical and methodological tools used in forest sensing using the UAVs are based on machine learning technology using the neural network method. The proposed integrated approach makes it possible to determine taxation indicators of plantings and volumes of aboveground phytomass in the automated mode (or with elements of automation). To achieve automation of the process, data of three categories (RGB, Lidar, hyperspectrum) were collected and artificial intelligence methods based on the ultra-precise neural network were applied.

The remote determination of the main forest stands' taxation indices and, accordingly, the volume of the phytomass of the key carbon pools will also make it possible to remotely determine the carbon storage of forest ecosystems. This will allow the remote monitoring of carbon storage over large areas.

The similarity of phytomass values obtained by ground-based and remote sensing methods was statistically confirmed: Pearson criterion χ^2 (0.05; 5) = 2.79. Consequently, we can reliably use the distance-based phytomass estimation technique to calculate the carbon storage of forest stands.

The use of the UAVs with appropriate hardware will make it possible to obtain the forest characteristics necessary for the determination of the carbon storage as the need arises: to determine baseline characteristics at the start of forest climate projects, when monitoring the dynamics of carbon storage, and so on. It is also possible to determine the characteristics of forests after selective logging and after fires of different intensities. Of great importance is the possibility of using the scientific idea of the work and the described methodology to assess and predict the condition of the trees and shrubs as ecological frameworks and complex phytomeliorants on lands of different target categories. The methodological approach we propose is easily compatible with current technologies in the field of conservation, protection, and reproduction of forests in the Central Forest Steppe Region in the East European Plain. This allows the approach to be successfully integrated into the general system of forest monitoring and the assessment of stocks of wood and other plant resources, and into environmental and economic management in different types of cultural and natural ecosystems and in different landscapes.

Author Contributions: Conceptualization, V.S., M.V.L. and N.D.; methodology, V.S. and D.L.; formal analysis, V.S., S.S., S.M. and M.V.L.; investigation, V.S., S.S., S.M. and M.V.L.; data processing and analysis, A.S.; writing—original draft preparation, V.S., D.L., S.M., M.V.L., A.S. and P.L.; writing—review and editing, V.S., D.L., S.S., S.M. and M.V.L.; organizational conceptualization and administrative support, V.S. and M.V.L.; participation in correction—V.S., M.V.L. and N.D.; visualization, V.S. All authors have read and agreed to the published version of the manuscript.

Funding: This research was funded the research was carried out within the state assignment of Ministry of Science and Higher Education of the Russian Federation (theme No. FZUR-2024-0002) "Development of a technological solutions aimed at increasing the efficiency of forest fire monitoring and detailed assessment of the consequences of forest fires in the conditions of the Central forest steppe"(State contract No. 075-03-2024-166). The research was supported by the Interdisciplinary Scientific and Educational School of Moscow University "Brain, Cognitive Systems, Artificial Intelligence".

Institutional Review Board Statement: Not applicable.

Informed Consent Statement: Not applicable.

Data Availability Statement: The data presented in this study are available on request from the corresponding author. The data are not publicly available due to privacy restrictions.

Conflicts of Interest: The authors declare no conflicts of interest.

References

1. Hoegh-Guldberg, O.; Jacob, D.; Taylor, M.; Guillen Bolanos, T.; Bindi, M.; Brown, S.; Camilloni, I.; Diedhiou, A.; Djalante, R.; Ebi, K.; et al. The human imperative of stabilizing global climate change at 1.5 °C. *Science* **2019**, *365*, eaaw6974. [CrossRef] [PubMed]
2. Sippel, S.; Meinshausen, N.; Fischer, E.M.; Szekely, E.; Knutti, R. Climate change now detectable from any single day of weather at global scale. *Nat. Clim. Change* **2020**, *10*, 35–41. [CrossRef]
3. Malysheva, N.V.; Moiseev, B.N.; Filipchuk, A.N.; Zolina, T.A. Methods for assessing the carbon balance in forest ecosystems and the possibility of their use for calculating annual carbon sequestration. *For. Bull.* **2017**, *21*, 4–13. [CrossRef]
4. Zhang, X.; Liu, D.; Ma, J.; Wang, X.; Li, Z.; Zheng, D. Visible Near-Infrared Hyperspectral Soil Organic Matter Prediction Based on Combinatorial Modeling. *Agronomy* **2024**, *14*, 789. [CrossRef]

5. Braga, C.I.; Crisan, V.E.; Petritan, I.C.; Scarlatescu, V.; Vasile, D.; Lazar, G.; Petritan, A.M. Short-Term Effects of Anthropogenic Disturbances on Stand Structure, Soil Properties, and Vegetation Diversity in a Former Virgin Mixed Forest. *Forests* **2023**, *14*, 742. [CrossRef]
6. Chhabra, A.; Palria, S.; Dadhwal, V.K. Spatial distribution of phytomass carbon in Indian forests. *Glob. Change Biol.* **2002**, *8*, 1230–1239. [CrossRef]
7. Chen, L.; Song, M.; Song, T.-Q.; Wang, H.; Zeng, F.-P.; Peng, W.-X.; Du, H.; Han, C.; Su, L. Carbon storage and its allocation in soft broadleaved forests with different stand ages in Guangxi. *Chin. J. Ecol.* **2017**, *36*, 592–600. [CrossRef]
8. Braga, C.I.; Petrea, S.; Radu, G.R.; Cucu, A.B.; Serban, T.; Zaharia, A.; Leca, S. Carbon Sequestration Dynamics in Peri-Urban Forests: Comparing Secondary Succession and Mature Stands under Varied Forest Management Practices. *Land* **2024**, *13*, 492. [CrossRef]
9. Zhou, Y.; She, J.; Zhu, X. Dynamic analysis of biodiversity, carbon storage and environmental factors of coniferous forest in Loudi City, Hunan Province. *Int. J. Low-Carbon Technol.* **2022**, *17*, 831–840. [CrossRef]
10. Zielonka, A.; Drewnik, M.; Musielok, Ł.; Dyderski, M.K.; Struzik, D.; Smułek, G.; Ostapowicz, K. Biotic and Abiotic Determinants of Soil Organic Matter Stock and Fine Root Biomass in Mountain Area Temperate Forests—Examples from Cambisols under European Beech, Norway Spruce, and Silver Fir (Carpathians, Central Europe). *Forests* **2021**, *12*, 823. [CrossRef]
11. Bisht, S.; Bargali, S.S.; Bargali, K.; Rawat, G.S.; Rawat, Y.S.; Fartyal, A. Influence of Anthropogenic Activities on Forest Carbon Stocks—A Case Study from Gori Valley, Western Himalaya. *Sustainability* **2022**, *14*, 16918. [CrossRef]
12. Diao, J.; Liu, J.; Zhu, Z.; Wei, X.; Li, M. Active forest management accelerates carbon storage in plantation forests in Lishui, southern China. *For. Ecosyst.* **2022**, *9*, 100004. [CrossRef]
13. Indriyani, L.; Bana, S.; Yasin, A.; Sudia, L.B.; Kahirun; Midi, L.O. Hardin The Potential of Blue Carbon Stocks and Carbon Dioxide Absorption in Mangrove Forests to Support Low Carbon Emission Development in Southeast Sulawesi Province, Indonesia. *Int. J. Adv. Sci. Eng. Inf. Technol.* **2020**, *10*, 2526–2535. [CrossRef]
14. Molotoks, A.; Stehfest, E.; Doelman, J.; Albanito, F.; Fitton, N.; Dawson, T.P.; Smith, P. Global projections of future cropland expansion to 2050 and direct impacts on biodiversity and carbon storage. *Glob. Change Biol.* **2018**, *24*, 5895–5908. [CrossRef] [PubMed]
15. Payne, N.J.; Cameron, D.A.; Leblanc, J.D.; Morrison, I.K. Carbon storage and net primary productivity in Canadian boreal mixedwood stands. *J. For. Res.* **2019**, *30*, 1667–1678. [CrossRef]
16. Tian, L.; Tao, Y.; Fu, W.X.; Li, T.; Ren, F.; Li, M.Y. Dynamic Simulation of Land Use/Cover Change and Assessment of Forest Ecosystem Carbon Storage under Climate Change Scenarios in Guangdong Province, China. *Remote Sens.* **2022**, *14*, 2330. [CrossRef]
17. Xiao, J.F.; Chevallier, F.; Gomez, C.; Guanter, L.; Hicke, J.A.; Huete, A.R.; Ichii, K.; Ni, W.J.; Pang, Y.; Rahman, A.F.; et al. Remote sensing of the terrestrial carbon cycle: A review of advances over 50 years. *Remote Sens. Environ.* **2019**, *233*, 111383. [CrossRef]
18. Strandberg, G.; Kjellström, E. Climate Impacts from Afforestation and Deforestation in Europe. *Earth Interact.* **2018**, *23*, 1–27. [CrossRef]
19. Veiga, J.; Lima, A.L.; Brito, A. Effects of deforestation at different spatial scales on the climate of the Amazon basin. *Clim. Res.* **2023**, *91*, 21. [CrossRef]
20. Volodkin, A.A.; Volodkina, O.A.; Larionov, M.V. Dynamics of reproduction of forest plantations in the forest-steppe zone of the Middle Volga Region. *IOP Conf. Ser. Earth Environ. Sci.* **2022**, *979*, 012101. [CrossRef]
21. Chmura, D.J.; Modrzynski, J. Sensitivity of height growth response to climate change does not vary with age in common garden among Norway spruce populations from elevational gradients. *For. Ecol. Manag.* **2023**, *542*, 121118. [CrossRef]
22. Sitková, Z.; Hlásny, T. Editorial to the thematic issue: Ecological interactions in Central European forest under climate change. *Cent. Eur. For. J.* **2020**, *66*, 189–190. [CrossRef]
23. Vedernikov, K.E.; Bukharina, I.L.; Udalov, D.N.; Pashkova, A.S.; Larionov, M.V.; Mazina, S.E.; Galieva, A.R. The State of Dark Coniferous Forests on the East European Plain Due to Climate Change. *Life* **2022**, *12*, 1874. [CrossRef] [PubMed]
24. Vinceti, B.; Manica, M.; Lauridsen, N.; Verkerk, P.J.; Lindner, M.; Fady, B. Managing forest genetic resources as a strategy to adapt forests to climate change: Perceptions of European forest owners and managers. *Eur. J. For. Res.* **2020**, *139*, 1107–1119. [CrossRef]
25. Gibadulina, I.I.; Larionov, M.V.; Maslennikova, N.N. Anatomical and morphological features of the leaves of *Tilia cordata* Mill. as an indicator of the adaptive capabilities of the species to the conditions of the urban environment. *IOP Conf. Ser. Earth Environ. Sci.* **2022**, *988*, 032082. [CrossRef]
26. Hruška, J.; Oulehle, F.; Chuman, T.; Kolář, T.; Rybníček, M.; Trnka, M.; McDowell, W.H. Forest growth responds more to air pollution than soil acidification. *PLoS ONE* **2023**, *18*, e0256976. [CrossRef]
27. Larionov, M.V.; Larionov, N.V.; Siraeva, I.S.; Gromova, T.S. Ecological and aesthetic significance of an auto-trophic component of artificial ecosystems in ensuring of the environmental comfort and the public health protection. *IOP Conf. Ser. Earth Environ. Sci.* **2020**, *421*, 082002. [CrossRef]
28. Larionov, M.V.; Soldatova, V.V.; Logacheva, E.A.; Larionov, N.V. An ecological analysis of the composition and condition of woody plants in urban and suburban ecosystems of the Khopyor River Region. *IOP Conf. Ser. Earth Environ. Sci.* **2020**, *421*, 062025. [CrossRef]

29. Larionov, M.V.; Tarakin, A.V.; Minakova, I.V.; Sentishcheva, E.A.; Bukreeva, T.N. Creation of artificial phytocenoses with controlled properties as a tool for managing cultural ecosystems and landscapes. *IOP Conf. Ser. Earth Environ. Sci.* **2021**, *848*, 012127. [CrossRef]
30. Bolte, A.; Lambertz, B.; Steinmeyer, A.; Kallweit, R.; Meesenburg, H. Zur Funktion der Bodenvegetation im Nährstoffhaushalt von Wäldern—Studien auf Dauerbeobachtungsflächen des EU Level II-Programms in Norddeutschland (The role of the herbaceous vegetation in forest nutrient cycling—Studies on permanent plots of the EU Level II program in North Germany). *Forstarchiv* **2004**, *75*, 207–220. [CrossRef]
31. Davis, Z.; Nesbitt, L.; Guhn, M.; van den Bosch, M. Assessing changes in urban vegetation using Normalised Difference Vegetation Index (NDVI) for epidemiological studies. *Urban For. Urban Green.* **2023**, *88*, 128080. [CrossRef]
32. Kaul, M.; Mohren, G.M.J.; Dadhwal, V.K. Phytomass carbon pool of trees and forests in India. *Clim. Change* **2011**, *108*, 243–259. [CrossRef]
33. Larionov, M.V.; Volodkin, A.A.; Volodkina, O.A.; Lebedev, E.V.; Khanbabayeva, O.E.; Tazina, S.V.; Kozlova, E.A.; Orlova, E.E.; Zubik, I.N.; Bogdanova, V.D.; et al. Features of the Territorial Distribution, Composition and Structure of Phytocenoses with the Participation of *Fraxinus excelsior*, Their Resource Qualities, Ecological and Economic Importance (Southeastern Part of the East European Plain). *Life* **2023**, *13*, 93. [CrossRef] [PubMed]
34. Lebedev, V.M.; Lebedev, E.V.; Sorokopudov, V.N.; Larionov, M.V. Root Nutrition, Photosynthesis, and Net Primary Production in Tree Stands of the genus Picea at the Organism Level within the Range in Russia. *Lesn. Zhurnal* **2023**, *1*, 38–50. [CrossRef]
35. Mayer, M.; Prescott, C.E.; Abaker, W.E.A.; Augusto, L.; Cécillon, L.; Ferreira, G.W.D.; James, J.; Jandl, R.; Katzensteiner, K.; Laclau, J.-P.; et al. Tamm Review: Influence of Forest Management Activities on Soil Organic Carbon Stocks: A Knowledge Synthesis. *For. Ecol. Manag.* **2020**, *466*, 118127. [CrossRef]
36. Li, D.R.; Wang, C.W.; Hu, Y.M.; Liu, S.G. General Review on Remote Sensing-Based Biomass Estimation. *Geomat. Inform. Sci. Wuhan Univ.* **2012**, *37*, 631–635. [CrossRef]
37. Slavskiy, V.A.; Mironenko, A.V.; Matveev, S.M.; Litovchenko, D.A. Taxation-interpretation indicators of plantations as the basis for the development of machine learning models for remote monitoring of forests. *Proc. St. Petersburg Res. Inst. For.* **2022**, *4*, 99–114. [CrossRef]
38. Ameray, A.; Bergeron, Y.; Valeria, O.; Montoro Girona, M.; Cavard, X. Forest Carbon Management: A Review of Silvicultural Practices and Management Strategies Across Boreal, Temperate and Tropical Forests. *Curr. For. Rep.* **2021**, *7*, 245–266. [CrossRef]
39. Wijedasa, L.S.; Jain, A.; Ziegler, A.D.; Evans, T.A.; Fung, T. Estimating Carbon Biomass in Forests Using Incomplete Data. *Biotropica* **2021**, *53*, 397–408. [CrossRef]
40. Shvidenko, A.Z.; Shepaschenko, D.G. Carbon budget of Russian forests. *Sib. For. J.* **2014**, *1*, 69–92. Available online: https://xn--80abmehbaibgnewcmzjeef0c.xn--p1ai/articles/archive/arkhiv-2014/uglerodnyy-byudzhet-lesov-rossii/ (accessed on 21 March 2024).
41. Schepaschenko, D.; McCallum, I.; Shvidenko, A.; Fritz, S.; Kraxner, F.; Obersteiner, M. A new hybrid land cover dataset for Russia: A methodology for integrating statistics, remote sensing and in-situ information. *J. Land Use Sci.* **2011**, *6*, 245–259. [CrossRef]
42. Pilli, R.; Alkama, R.; Cescatti, A.; Kurz, W.A.; Grassi, G. The European Forest Carbon Budget under Future Climate Conditions and Current Management Practices. *Biogeosciences* **2022**, *19*, 3263–3284. [CrossRef]
43. Liang, L.; Geng, D.; Yan, J.; Qiu, S.; Shi, Y.; Wang, S.; Wang, L.; Zhang, L.; Kang, J. Remote Sensing Estimation and Spatiotemporal Pattern Analysis of Terrestrial Net Ecosystem Productivity in China. *Remote Sens.* **2022**, *14*, 1902. [CrossRef]
44. Heath, L.; Nichols, M.; Smith, J.; Mills, J. *FORCARB2. An Updated Version of U.S. Forest Carbon Budget Model*; General Technical Report; NRS-67 USDA Forest Service, Northern Research Station: Newtown Square, PA, USA, 2010; 52p. Available online: http://nrs.fs.fed.us/pubs/35613 (accessed on 21 March 2024).
45. Liu, S.; Bond-Lamberty, B.; Hicke, J.A. Simulation the impacts of disturbances on forest carbon cycling in North America: Processes, data, models and challenges. *J. Geophys. Res.* **2011**, *116*, 22. [CrossRef]
46. Ma, L.; Hurtt, G.; Tang, H.; Lamb, R.; Lister, A.; Chini, L.; Dubayah, R.; Armston, J.; Campbell, E.; Duncanson, L.; et al. Spatial heterogeneity of global forest aboveground carbon stocks and fluxes constrained by spaceborne lidar data and mechanistic modeling. *Glob. Change Biol.* **2023**, *29*, 3378–3394. [CrossRef] [PubMed]
47. Vahedi, A.A. Monitoring soil carbon pool in the Hyrcanian coastal plain forest of Iran: Artificial neural network application in comparison with developing traditional models. *CATENA* **2017**, *152*, 182–189. [CrossRef]
48. Yu, Y.; Guo, B.; Wang, C.; Zang, W.; Huang, X.; Wu, Z.; Xu, M.; Zhou, K.; Li, J.; Yang, Y. Carbon storage simulation and analysis in Beijing-Tianjin-Hebei region based on CA-plus model under dual-carbon background. *Geomat. Nat. Hazards Risk* **2023**, *14*, 1. [CrossRef]
49. Nyamukuru, A.; Whitney, C.; Tabuti, J.R.S.; Esaete, J.; Low, M. Allometric models for aboveground biomass estimation of small trees and shrubs in African savanna ecosystems. *Trees For. People* **2023**, *11*, 100377. [CrossRef]
50. Sarker, M.L.R.; Nichol, J.; Iz, H.B.; Bin Ahmad, B.; Rahman, A.A. Forest Biomass Estimation Using Texture Measurements of High-Resolution Dual-Polarization C-Band SAR Data. *IEEE Trans. Geosci. Remote* **2013**, *51*, 3371–3384. [CrossRef]
51. Ali, A.; Lin, S.L.; He, J.K.; Kong, F.M.; Yu, J.H.; Jiang, H.S. Climate and soils determine aboveground biomass indirectly via species diversity and stand structural complexity in tropical forests. *For. Ecol. Manag.* **2019**, *432*, 823–831. [CrossRef]

52. Strategy for the Development of the Forest Complex of the Russian Federation until 2023 (Decree of the Government of the Russian Federation dated 11 February 2021 No. 312-r). Available online: https://docs.cntd.ru/document/573658653 (accessed on 1 September 2023).
53. Deng, L.; Liu, S.G.; Kim, D.G.; Peng, C.H.; Sweeney, S.; Shangguan, Z.P. Past and future carbon sequestration benefits of China's grain for green program. *Glob. Environ. Change* **2017**, *47*, 13–20. [CrossRef]
54. Gong, W.; Sun, J.; Shi, S.; Yang, J.; Du, L.; Zhu, B.; Song, S. Investigating the Potential of Using the Spatial and Spectral Information of Multispectral LiDAR for Object Classification. *Sensors* **2015**, *15*, 21989–22002. [CrossRef] [PubMed]
55. Slavskiy, V.A.; Matveev, S.M. Some aspects of the laying of permanent trial plots during the state inventory of forests. *Lesotechnical J.* **2021**, *11*, 56–63. [CrossRef] [PubMed]
56. Huang, H.B.; Liu, C.X.; Wang, X.Y.; Zhou, X.L.; Gong, P. Integration of multi-resource remotely sensed data and allometric models for forest aboveground biomass estimation in China. *Remote Sens. Environ.* **2019**, *221*, 225–234. [CrossRef]
57. Zhang, R.; Zhou, X.H.; Ouyang, Z.T.; Avitabile, V.; Qi, J.G.; Chen, J.Q.; Giannico, V. Estimating aboveground biomass in subtropical forests of China by integrating multisource remote sensing and ground data. *Remote Sens. Environ.* **2019**, *232*, 111341. [CrossRef]
58. Zhang, Y.Z.; Liang, S.L.; Yang, L. A Review of Regional and Global Gridded Forest Biomass Datasets. *Remote Sens.* **2019**, *11*, 2744. [CrossRef]
59. Chopping, M.; Wang, Z.S.; Schaaf, C.; Bull, M.A.; Duchesne, R.R. Forest aboveground biomass in the southwestern United States from a MISR multi-angle index, 2000–2015. *Remote Sens. Environ.* **2022**, *275*, 112964. [CrossRef]
60. Li, Y.; Xiao, J.; Cong, N.; Yu, X.; Lin, Y.; Liu, T.; Qi, G.; Ren, P. Modeling Ecological Resilience of Alpine Forest under Climate Change in Western Sichuan. *Forests* **2023**, *14*, 1769. [CrossRef]
61. Zhang, Z.; Tian, X.; Chen, R.X.; He, Q.S. Review of methods on estimating forest above ground biomass. *J. Beijing For. Univ.* **2011**, *33*, 144–150. Available online: https://www.researchgate.net/publication/286837566_Review_of_methods_on_estimating_forest_above_ground_biomass (accessed on 21 March 2024).
62. Demina, N.A.; Karpov, A.A.; Voronin, V.V.; Lopatin, E.V.; Bogdanov, A.P. Possible use of remote sensing for reforestation processes in Arctic zone of European Russia. *Arct. Environ. Res.* **2018**, *18*, 106–113. [CrossRef]
63. Malysheva, N.V.; Zolina, T.A.; Dedova, V.Y. Estimation of carbon accumulation by Russian forests: Geospatial aspect. *InterCarto InterGIS* **2017**, *1*, 373–382. [CrossRef]
64. Filipchuk, A.N.; Moiseev, B.N.; Malysheva, N.V. Methodology for taking into account the absorption of CO_2 by forests in Russia. In Proceedings of the 2nd International Scientific and Technical Conference "Forests of Russia: Politics, Industry, Science, Education", St. Petersburg, Russia, 24–26 May 2017; Volume 2, pp. 155–158.
65. Baccini, A.; Walker, W.; Carvalho, L.; Farina, M.; Sulla-Menashe, D.; Houghton, R.A. Tropical forests are a net carbon source based on aboveground measurements of gain and loss. *Science* **2017**, *358*, 230–233. [CrossRef] [PubMed]
66. Beaudoin, A.; Bernier, P.Y.; Guindon, L.; Villemaire, P.; Guo, X.J.; Stinson, G.; Bergeron, T.; Magnussen, S.; Hall, R.J. Mapping attributes of Canada's forests at moderate resolution through kNN and MODIS imagery. *Can. J. For. Res.* **2014**, *44*, 521–532. [CrossRef]
67. Dube, T.; Mutanga, O. Evaluating the utility of the medium-spatial resolution Landsat 8 multispectral sensor in quantifying aboveground biomass in uMgeni catchment, South Africa. *ISPRS J. Photogramm.* **2015**, *101*, 36–46. [CrossRef]
68. Powell, S.L.; Cohen, W.B.; Healey, S.P.; Kennedy, R.E.; Moisen, G.G.; Pierce, K.B.; Ohmann, J.L. Quantification of live aboveground forest biomass dynamics with Landsat time-series and field inventory data: A comparison of empirical modeling approaches. *Remote Sens. Environ.* **2010**, *114*, 1053–1068. [CrossRef]
69. Fremout, T.; Vinatea, J.C.D.; Thomas, E.; Huaman-Zambrano, W.; Salazar-Villegas, M.; de la Fuente, D.L.; Bernardino, P.N.; Atkinson, R.; Csaplovics, E.; Muys, B. Site-specific scaling of remote sensing-based estimates of woody cover and aboveground biomass for mapping long-term tropical dry forest degradation status. *Remote Sens. Environ.* **2022**, *276*, 113040. [CrossRef]
70. Fan, W.Y.; Li, M.Z.; Yang, J.M. Forest Biomass Estimation Models of Remote Sensing in Changbai Mountain Forests. *Sci. Silvae Sinicae* **2011**, *47*, 16–20. [CrossRef]
71. Hopkinson, C.; Chasmer, L.; Gynan, C.; Mahoney, C.; Sitar, M. Multisensor and Multispectral LiDAR Characterization and Classification of a Forest Environment. *Can. J. Remote Sens.* **2016**, *42*, 501–520. [CrossRef]
72. Zou, X.; Zhao, G.; Li, J.; Yang, Y.; Fang, Y. 3D Land Cover Classification Based on Multispectral Lidar Point Clouds. *Int. Arch. Photogramm. Remote Sens. Spatial Inf. Sci.* **2016**, *XLI-B1*, 741–747. [CrossRef]
73. Kukkonen, M.; Maltamo, M.; Korhonen, L.; Packalen, P. Multispectral Airborne LiDAR Data in the Prediction of Boreal Tree Species Composition. *IEEE Trans. Geosci. Remote Sens.* **2019**, *57*, 3462–3471. [CrossRef]
74. Dalponte, M.; Ene, L.T.; Gobakken, T.; Næsset, E.; Gianelle, D. Prediction of Forest Attributes with Multispectral Lidar Data. In Proceedings of the IGARSS 2018—2018 IEEE International Geoscience and Remote Sensing. Symposium, Valencia, Spain, 22–27 July 2018; pp. 7528–7531. [CrossRef]
75. Dalla Corte, A.P.; Souza, D.V.; Rex, F.E.; Sanquetta, C.R.; Mohan, M.; Silva, C.A.; Almeyda-Zambrano, A.M.; Prata, G.; de Almeida, D.R.; Trautenmüller, J.W.; et al. Forest inventory with high-density UAV-Lidar: Machine learning approaches for predicting individual tree attributes. *Comput. Electron. Agric.* **2020**, *179*, 105815. [CrossRef]
76. Immitzera, M.; Stepperb, C.; Böcka, S.; Straubb, C.; Atzbergera, C. Use of WorldView-2 stereo imagery and National Forest Inventory data for wall-to-wall mapping of growing stock. *For. Ecol. Manag.* **2016**, *359*, 232–246. [CrossRef]

77. Shang, C.; Coops, N.C.; Wulder, M.A.; White, J.C.; Hermosilla, T. Update and spatial extension of strategic forest inventories using time series remote sensing and modeling. *Int. J. Appl. Earth Obs. Geoinf.* **2020**, *84*, 1229–1241. [CrossRef]
78. Liang, X.; Hyyppä, J.; Kaartinen, H.; Lehtomäki, M.; Pyörälä, J.; Pfeifer, N.; Holopainen, M.; Brolly, G.; Pirotti, F.; Hackenberg, J.; et al. International benchmarking of terrestrial laser scanning approaches for forest inventories. *ISPRS J. Photogramm. Remote Sens.* **2018**, *144*, 137–179. [CrossRef]
79. Noordermeer, L.; Gobakken, T.; Næsset, E.; Bollandsas, O.-M. Predicting and mapping site index in operational forest inventories using bitemporal airborne laser scanner data. *For. Ecol. Manag.* **2020**, *457*, 117768. [CrossRef]
80. Lu, D.S.; Chen, Q.; Wang, G.X.; Liu, L.J.; Li, G.Y.; Moran, E. A survey of remote sensing-based aboveground biomass estimation methods in forest ecosystems. *Int. J. Digit. Earth.* **2016**, *9*, 63–105. [CrossRef]
81. Morsy, S.; Shaker, A.; El-Rabbany, A.; LaRocque, P.E. Airborne multispectral lidar data for land-cover classification and land/water mapping using different spectral indexes. *ISPRS Ann. Photogramm. Remote Sens. Spat. Inf. Sci.* **2016**, *3*, 217–224. [CrossRef]
82. Budei, B.C.; St-Onge, B.; Hopkinson, C.; Audet, F.A. Identifying the genus or species of individual trees using a three-wavelength airborne lidar system. *Remote Sens. Environ.* **2018**, *204*, 632–647. [CrossRef]
83. Ekhtari, N.; Glennie, C.; Fernandez-Diaz, J.C. Classification of Airborne Multispectral Lidar Point Clouds for Land Cover Mapping. *IEEE J. Sel. Top. Appl. Earth Obs. Remote Sens.* **2018**, *11*, 2068–2078. [CrossRef]
84. Wichmann, V.; Bremer, M.; Lindenberger, J.; Rutzinger, M.; Georges, C.; Petrini-Monteferri, F. Evaluating the potential of multispectral airborne lidar for topographic mapping and land cover classification. *ISPRS Ann. Photogram. Remote Sens. Spat. Inform. Sci.* **2015**, *3*, 113–119. [CrossRef]
85. Shokurov, A.V. System for remote size-adaptive variable detail raster image fragment retriever. *Probl. Inform.* **2012**, *1*, 25–39. Available online: https://cyberleninka.ru/article/n/sistema-udalennogo-izvlecheniya-fragmenta-rastrovogo-izobrazheniya-izmenyaemoy-detalizatsii-adaptirovannaya-k-ego-razmeru (accessed on 21 March 2024).
86. Shokurov, A.V. Irregularly Detailed Raster Image Coding Software Design with Respect to Adaptive and Interactive Image Analysis. *Softw. Eng.* **2011**, *5*, 17–28. Available online: http://iipo.tu-bryansk.ru/fileadmin/user_upload/_content_admins/Journal_PrI/ZHurnal_Programmnaja_Inzhenerija_2011_No_5.pdf (accessed on 21 March 2024).
87. Rules of Sanitary Safety in Forests: Decree of the Government of the Russian Federation of 9 December 2020 No. 2047. On the Rules of Sanitary Safety in Forests//Collection of Legislation. 2020. Available online: http://docs.cntd.ru/document/436736467 (accessed on 1 September 2023).
88. Lukina, N.V.; Tikhonova, E.V.; Danilova, M.A.; Bakhmet, O.N.; Kryshen, A.M.; Tebenkova, D.N.; Kuznetsova, A.I.; Smirnov, V.E.; Braslavskaya, T.Y.; Gornov, A.V.; et al. Associations between forest vegetation and the fertility of soil organic horizons in northwestern Russia. *For. Ecosyst.* **2019**, *6*, 34. [CrossRef]
89. Order of the Federal Forestry Agency. On Approval of Guidelines for the State Forest Inventory. 10 November 2011. Available online: http://docs.cntd.ru/document/436736785 (accessed on 3 September 2023).
90. Slavskiy, V.; Litovchenko, D.; Matveev, S.; Sheshnitsan, S.; Larionov, M.V. Assessment of Biological and Environmental Factors Influence on Fire Hazard in Pine Forests: A Case Study in Central Forest-Steppe of the East European Plain. *Land* **2023**, *12*, 103. [CrossRef]
91. Order of the Ministry of Natural Resources and Ecology of the Russian Federation dated December 29, 2021 No. 1024 "On Approval of the Rules for Reforestation, the Form, Composition, Procedure for Approving a Reforestation Project, Grounds for Refusing to Approve it, as well as Requirements for the Format in the Electronic Form of a Reforestation Project". Available online: https://docs.cntd.ru/document/728111110 (accessed on 4 September 2023).
92. On Approval of the Forest Management Instruction/Order of the Ministry of Natural Resources of Russia (Ministry of Natural Resources and Ecology of the Russian Federation) dated August 05 2022 No. 510. Available online: http://publication.pravo.gov.ru/Document/View/0001202209300058 (accessed on 4 September 2023).
93. Usoltsev, V.A.; Shobairi, S.O.R.; Chasovskikh, V.P. Comparing of Allometric Models of Single-tree Biomass Intended for Airborne Laser Sensing and Terrestrial Taxation of Carbon Pool in the Forests of Eurasia. *Nat. Resour. Model.* **2019**, *32*, e12187. [CrossRef]
94. Utkin, A.I.; Zamolodchikov, D.G.; Gulbe, T.A.; Gulbe, Y.A. Allometric equations for phytomass based on the data on pine, spruce, birch and aspen trees in European Russia. *Lesovedenie* **1996**, *6*, 36–46. Available online: https://elibrary.ru/item.asp?id=21266850 (accessed on 21 March 2024).
95. STATISTICA (13.0-StatSoft, 2021). Available online: https://statsoftstatistica.ru/ (accessed on 4 September 2023).
96. Tian, X.; Yan, M.; van der Tol, C.; Li, Z.; Su, Z.B.; Chen, E.X.; Li, X.; Li, L.H.; Wang, X.F.; Pan, X.D.; et al. Modeling forest above-ground biomass dynamics using multi-source data and incorporated models: A case study over the qilian mountains. *Agric. For. Meteorol.* **2017**, *246*, 1–14. [CrossRef]
97. Mehtatalo, L.; Lappi, J. *Biometry for Forestry and Environmental Data: With Examples in R*, 1st ed.; Chapman and Hall/CRC: New York, NY, USA, 2020. [CrossRef]
98. Lüdecke, D.; Makowski, D.; Waggoner, P.; Patil, I. *Assessment of Regression Models Performance*; CRAN: Indianapolis, IN, USA, 2020. [CrossRef]
99. Wang, J.; Yan, M.; Huang, Q.; Huang, R.; Zheng, Q. Diameter Distribution of Semi-natural Mixed Forest of Pinus massoniana and Broadleaved Trees Based on Stratification. *For. Res.* **2021**, *34*, 72–80. [CrossRef]

100. Yu, C.; Mao, J.-H.; Huang, X.; Zhang, Y. Carbon capture by phytomass storage and trading to mitigate climate change and preserve natural resources. *Environ. Sustain. Indic.* **2024**, *22*, 100358. [CrossRef]
101. GOST R 59328-2021 Topographic Aerial Photography. "Technical Requirements". Standartinform. 2021. Available online: https://rkc56.ru/attach/orenburg/docs/kodeks/gost_r_59328-2021-aerofotosemka.pdf (accessed on 4 September 2023).
102. Yang, B.; Zhang, W.; Lu, Y.; Zhang, W.; Wang, Y. Carbon Storage Dynamics of Secondary Forest Succession in the Central Loess Plateau of China. *Forests* **2019**, *10*, 342. [CrossRef]
103. Fatehi, P.; Damm, A.; Schaepman, M.E.; Kneubuhler, M. Estimation of Alpine Forest Structural Variables from Imaging Spectrometer Data. *Remote Sens.* **2015**, *7*, 16315–16338. [CrossRef]
104. Zhang, G.; Ganguly, S.; Nemani, R.R.; White, M.A.; Milesi, C.; Hashimoto, H.; Wang, W.L.; Saatchi, S.; Yu, Y.F.; Myneni, R.B. Estimation of forest aboveground biomass in California using canopy height and leaf area index estimated from satellite data. *Remote Sens. Environ.* **2014**, *151*, 44–56. Available online: https://scholar.google.com/scholar_lookup?title=Estimation+of+forest+aboveground+biomass+in+California+using+canopy+height+and+leaf+area+index+estimated+from+satellite+data&author=Zhang,+G.&author=Ganguly,+S.&author=Nemani,+R.R.&author=White,+M.A.&author=Milesi,+C.&author=Hashimoto,+H.&author=Wang,+W.L.&author=Saatchi,+S.&author=Yu,+Y.F.&author=Myneni,+R.B.&publication_year=2014&journal=Remote+Sens.+Environ.&volume=151&pages=44%E2%80%9356&doi=10.1016/j.rse.2014.01.025 (accessed on 10 February 2024). [CrossRef]
105. Levick, S.R.; Shendryk, Y.; Setterfield, S.A.; Rossiter-Rachor, N.A. *Evaluation of Satellite Remote Sensing Pathways for Mapping and Monitoring of Gamba Grass for the Savanna Fire Management Methodology*; CSIRO and Charles Darwin University: Darwin, Australia, 2018; Volume 20, 27p. Available online: https://nesplandscapes.edu.au/wp-content/uploads/2019/10/Evaulation-of-remote-sensing-for-mapping-monitoring-gamba.pdf (accessed on 10 February 2024).
106. Hill, S.; Latifi, H.; Heurich, M.; Müller, J. Individual-tree- and stand-based development following natural disturbance in a heterogeneously structured forest: A LiDAR-based approach. *Ecol. Inform.* **2017**, *38*, 12–25. [CrossRef]
107. Marselis, S. *Mapping Forest Fuel Structure with LiDAR*; Australian National University Press: Canberra, Australia, 2014; pp. 1–35.
108. Pierfranco Costabile, P.; Carmelina Costanzo, C.; Gianluca De Lorenzo, G.; De Santis, R.; Penna, N.; Macchione, F. Terrestrial and airborne laser scanning and 2-D modelling for 3-D flood hazard maps in urban areas: New opportunities and perspectives. *Environ. Model. Softw.* **2021**, *135*, 104889. [CrossRef]
109. Patil, P.; Dutta, D.; Biradar, C.; Singh, M. Quantification of the terrestrial phytomass and carbon in the mountainous forest ecosystem using remote sensing and in-situ observations. *Int. Arch. Photogramm. Remote Sens. Spatial Inf. Sci.* **2015**, *XL-7/W3*, 483–487. [CrossRef]
110. Heyojoo, B.P.; Nandy, S. Estimation of above-ground phytomass and carbon in tree resources outside the forest (TROF): A geo-spatial approach. *Banko Janakari* **2015**, *24*, 34–40. [CrossRef]
111. Tian, L.; Wu, X.; Tao, Y.; Li, M.; Qian, C.; Liao, L.; Fu, W. Review of Remote Sensing-Based Methods for Forest Aboveground Biomass Estimation: Progress, Challenges, and Prospects. *Forests* **2023**, *14*, 1086. [CrossRef]
112. Reinmann, A.B.; Smith, I.A.; Thompson, J.R.; Hutyra, L.R. Urbanization and fragmentation mediate temperate forest carbon cycle response to climate. *Environ. Res. Lett.* **2020**, *15*, 114036. [CrossRef]
113. Zhao, J.; Liu, D.; Zhu, Y.; Peng, H.; Xie, H. A review of forest carbon cycle models on spatiotemporal scales. *J. Clean. Prod.* **2022**, *339*, 130692. [CrossRef]
114. Liu, D.; Yan, H.; Yu, C.L.; Yin, S.P.; Wang, C.W.; Gong, L.J. Quantitative Assessment of Climate Change Impacts on Forest Ecosystems Get access Arrow. *For. Sci.* **2022**, *69*, 143–157. [CrossRef]
115. Singh, K.; Sharma, P.C. Above-ground tree outside forest (TOF) phytomass and carbon estimation in the semi-arid region of southern Haryana: A synthesis approach of remote sensing and field data. *J. Earth Syst. Sci.* **2013**, *121*, 1469–1482. [CrossRef]
116. Suthari, S.; Raju, V.S.; Singh, S. An assessment of the aboveground phytomass and carbon levels of the forests of northern Telangana, India, using a geospatial technique. *Biodiversity* **2020**, *21*, 227–237. [CrossRef]
117. Mudrák, O.; Angst, Š.; Angst, G.; Veselá, H.; Schnablová, R.; Herben, T.; Frouz, J. Ecological significance of standing dead phytomass: Marcescence as a puzzle piece to the nutrient cycle in temperate ecosystems. *J. Ecol.* **2023**, *111*, 2245–2256. [CrossRef]
118. Nandy, S.; Ghosh, S.; Kushwaha, S.P.S.; Senthil Kumar, A. Remote Sensing-Based Forest Biomass Assessment in Northwest Himalayan Landscape. In *Remote Sensing of Northwest Himalayan Ecosystems*; Navalgund, R., Kumar, A., Nandy, S., Eds.; Springer: Singapore, 2017. [CrossRef]
119. Schiffman, P.M.; Johnson, W.C. Phytomass and detrital carbon storage during forest regrowth in the southeastern United States Piedmont. *Can. J. For. Res.* **2011**, *19*, 69–78. [CrossRef]
120. Hetemäki, L.; Jyrki, K.; Peltola, H. (Eds.) *Forest Bioeconomy and Climate Change*; Series: Managing Forest Ecosystems, MAFE; Springer Nature Switzerland AG: Cham, Switzerland, 2022; Volume 42, 257p. [CrossRef]
121. Ho, K.V.; Kröel-Dulay, G.; Tölgyesi, C.; Bátori, Z.; Tanács, E.; Kertész, M.; Török, P.; Erdős, L. Non-native tree plantations are weak substitutes for near-natural forests regarding plant diversity and ecological value. *For. Ecol. Manag.* **2023**, *531*, 120789. [CrossRef]

122. Sena, K.N.; Maltoni, K.; Troleis, M.J.B.; Faria, G.A. Forest harvest management systems and residual phytomass affecting physical properties of a sandy soil. *Rev. Bras. de Ciência do Solo* **2021**, *45*, e0200190. [CrossRef]
123. Zhao, H.; Zhao, D.; Jiang, X.; Zhang, S.; Lin, Z. Assessment of Urban Forest Ecological Benefit Based on the i-Tree Eco Model—A Case Study of Changchun Central City. *Forests* **2023**, *14*, 1304. [CrossRef]

life

Article

Defense Mechanisms of *Xylopia aromatica* (Lam.) Mart. in the Dry Season in the Brazilian Savanna

Felipe Campos [1,*], Maria Vieira [1], Marília Sousa [1], Letícia Jorge [1], Gisela Ferreira [1], Marcia Marques [2] and Carmen Boaro [1]

[1] Biodiversity and Biostatistics Departament, Biosciences Institute, São Paulo State University (UNESP), Campus (Botucatu), P.O. Box 510, Botucatu 18618-970, SP, Brazil; maria.vieira@unesp.br (M.V.); marilia.caixetas@gmail.com (M.S.); leticia.g.jorge@unesp.br (L.J.); gisela.ferreira@unesp.br (G.F.); carmen.boaro@unesp.br (C.B.)

[2] Agronomic Institute of Campinas (IAC), Plant Genetic Resources Center, Campinas 13075-630, SP, Brazil; marcia.marques@sp.gov.br

* Correspondence: felipe.girotto@unesp.br; +55-14-3880-0124

Abstract: Water availability and light during the dry and rainy seasons in the Cerrado may influence plants' stomatal movement and the entry of CO_2 for organic synthesis, which is the main electron drain. A lower stomatal conductance may contribute to the energy accumulated in the chloroplasts being directed towards the synthesis of compounds, which contributes to the activity of antioxidant enzymes to neutralize reactive oxygen species. *Xylopia aromatica* is a characteristic Cerrado species, and it is often recommended for recovering degraded areas. This study aimed to investigate the influence of the dry and rainy seasons on the metabolic adjustments of *Xylopia aromatica* in a portion of the Brazilian savanna in the state of São Paulo. In the rainy season, better photosynthetic performance led to greater investment in essential oil production. In the dry season, the plants may direct part of their reducing sugars to the syntheses of carotenoids and anthocyanins, which may help the antioxidant enzymes to neutralize reactive oxygen species. Carotenoids assist in the dissipation of photosystem energy, which has the potential to cause oxidative stress. During this season, lower stomatal conductance prevented excessive water loss. These results suggest the acclimatization of this species to the conditions of the Brazilian savanna.

Keywords: Annonaceae; bicyclogermacrene; β-phellandrene; spathulenol; photosynthesis; antioxidant enzymes

1. Introduction

The Cerrado is a dry tropical forest in South America that covers a quarter of the national territory of Brazil and exhibits the greatest biodiversity in the world [1–4]. In recent decades, human actions, such as agricultural expansion and burning, have been the main causes of the degradation and fragmentation of the Brazilian savanna, resulting in the loss of around 50% of the original vegetation [1,5]. Therefore, studies involving local species are relevant to conserving the biodiversity of this Brazilian biome. It is particularly important to study the defense mechanisms used by species to survive in dry conditions. In this sense, *Xylopia aromatica* (Lam.) Mart. (Annonaceae) is a characteristic Brazilian savanna species, mainly recommended for the recovery of degraded areas of the Cerrado (SP) [6], Brazil (Resolução SMA nº 47/2003). *Xylopia aromatica* is also found in deciduous seasonal forests and the Amazonian savanna [6], being popularly known as pimenta-de-macaco, embira, pindaíba, and pimenta-de-bugre [7,8].

Climate change can cause variations in the dynamics of the Cerrado in both the dry and rainy seasons [1]. Thus, studying the interaction between plants and their changing environment contributes to understanding the impact on plant metabolism, helping to develop strategies for reforestation and preservation.

Xylopia aromatica seedlings introduced into a degraded area of the Cerrado as part of recovery efforts may present comparable nutritional status to those evaluated in preserved areas of the Cerrado. In addition, *Xylopia aromatica* has shown tolerance to high concentrations of leaf Fe and is therefore indicated for the recovery of areas contaminated with high levels of iron [9].

The metabolism of plant species in degraded environments can present, as a defense and acclimatization strategy, variations in both primary and specialized metabolism. Plants produce substances that belong to several chemical classes, which play important roles in their survival and adaptation to the ecosystem and contribute to their protection against abiotic and biotic stresses [10–13].

Phytochemical studies have shown that *Xylopia* contains several classes of substances that belong to specialized metabolic processes, including terpenes, amides, lignoids, alkaloids, and acetogenins [8,12,14,15]. Many of these substances have biological activity [16–18], including in humans, which may present variations in the different organs of the plant and with environmental conditions.

Among the terpenes, the monoterpenes α- and β-pinene and limonene, and the sesquiterpenes bicyclogermacrene and spathulenol, have been reported in the chemical profiles of essential oils extracted from the leaves, bark, flowers, fruits, and seeds of *Xylopia* plants in the Cerrado and Amazonian savanna [6,7,14,19–22].

Several studies have investigated the biological activity of *Xylopia aromatica* against biotic factors, such as fungi and bacteria. Nascimento et al. [21] evaluated the chemical profile and the antimicrobial and antifungal activity of the essential oils in the flowers and leaves of *Xylopia aromatica* collected in the Cerrado of Goiás, Brazil. The major compounds found in the leaves were spathulenol, khusinol, and bicyclogermacrene, and those found in the flowers included pentadecane-2-one, bicyclogermacrene, 7-epi-α-eudesmol, and khusinol. The essential oils of leaves and flowers showed lower minimum inhibitory concentrations (MIC) for *Streptococcus pyogenes* (200 and 100 $\mu g\ mL^{-1}$, respectively). The essential oil in the leaves showed moderate inhibitory activity (500 $\mu g\ mL^{-1}$) for *Candida albicans*. The essential oil obtained from the leaves of *Xylopia aromatica* of the Amazonian savanna, Amazonas, and Brazil showed a predominance of spathulenol, *trans*-pinocarveol, and dihydrocarveol, with strong activity against *Streptococcus sanguinis* [22]. Essential oil from *Xylopia aromatica* leaves collected in the Cerrado of Botucatu, SP, Brazil, contained α-pinene and β-pinene as major substances and showed antifungal activity against *Candida* and *Cryptococcus* species [23].

Another aspect of extreme importance, which has been less studied, is the effect the environment can have on the levels of enzymes that allow a plant to survive and regulate its primary and specialized metabolisms. Plants are sessile organisms subject to environmental conditions that can generate oxidative stress. This stress can be caused by variations in rainfall intensity, temperature, light, competition with other plants, pests, diseases, and herbivores. Under stress conditions, reactive oxygen species (ROS), molecules formed during normal metabolic functions in chloroplasts, mitochondria, and peroxisomes [24], are generated by the electron transport system, causing cellular damage or, depending on concentration, acting as signaling molecules that exert multiple defense responses. In these conditions, the plants depend on antioxidants for ROS neutralization and defense.

Antioxidant enzymes, including superoxide dismutase and peroxidases, can reverse oxidative stress [25]. These enzymes and the level of lipid peroxidation can be used [26] to monitor the damage caused by ROS and the malondialdehyde accumulation [27], acting as molecular markers to determine the stress level in plants.

The variation in water availability and light incidence in the Cerrado can influence the stomatal movement, as well as the entry of carbonic gas for organic synthesis, the main electron drain. A lower stomatal conductance may contribute to the accumulation of reducing agents in chloroplasts, which can be used in the synthesis of compounds that participate in specialized metabolism. This characterizes the non-enzymatic antioxidant system, which contributes to the activity of antioxidant enzymes to neutralize reactive oxygen species [28], such as mono- and sesquiterpenes. These substances are produced from the union of two or three isoprenes, respectively. The formation of isoprene requires a skeleton with five carbons, three ATP molecules, the reducing agent $NAD(P)H^+$, and an electron donor source for the reduction of methylerythritol phosphate into isoprene. This isoprene route can also play a role in the dissipation of excess photosynthetic energy [28–30] and constitutes an important protection mechanism when the plant is subjected to a high light incidence. Under these conditions, drought or flooding stress may involve the emission of terpenes, which function as stress markers (as occurs in plant–plant and plant–insect interactions [31–33]) and act to stimulate a mechanism to overcome the stress factor.

No studies investigating the association between water availability and the defense mechanisms of the primary and specialized metabolisms of species native to the Cerrado have been found in the literature. This study aimed to investigate the influence of dry and rainy seasons on the metabolic adjustments in the photosynthetic and terpene profile of *Xylopia aromatica* in the fragment of Brazilian savanna in the state of São Paulo, Brazil.

2. Materials and Methods

2.1. Plant Material and Environmental Conditions

Twenty-eight individual *Xylopia aromatica* plants were evaluated. Fourteen were collected during the dry seasons, Dry1 (September 2016) and Dry2 (August 2017), and fourteen were collected during the rainy seasons, Rainy1 (February 2017) and Rainy2 (February 2018).

The evaluations were conducted on plants obtained from the Cerrado remnant at Rio Bonito, located in the municipality of Botucatu, state of São Paulo, Brazilian Southeastern region (geographical coordinates 22°42′09.4″ S, 48°20′64.1″ W and altitude of 514 m).

Reproductive materials from the species were collected, exsiccated, and deposited at the Herbarium Irina Delanova Gemtchujnicov (BOTU), located in the Biosciences Institute of Botucatu, UNESP, voucher n° 32478.

The chemical and physical characteristics of the collected soil samples were evaluated, and a pool consisting of 10 repetitions (20–40 cm) was made in the study area at each collection site. The chemical and physical characteristics of the soil showed differences in different seasons (Supplementary Table S1).

Measurements of precipitation (mm), relative humidity (%), and the average maximum and minimum atmospheric temperatures (°C) were obtained from the meteorological mini station (model PC400, Campbell Scientific, Shepshed, Loughborough, UK) located at the Unesp Biosciences Institute, Botucatu. The evaluations were carried out in the dry and rainy seasons, when the precipitation was lower and higher, respectively (Supplementary Figure S1).

2.2. Study Variables

The chlorophyll *a* fluorescence, gas exchange, leaf water potential, relative leaf water content, and yield and chemical profile of the essential oil were evaluated in 14 individuals in the Dry1 and Dry2 groups and another 14 individuals in the Rainy1 and Rainy2 groups.

Photosynthetic pigments, carbohydrates, antioxidant enzymes, and lipid peroxidation were evaluated in the Dry2 group and the Rainy2 group.

2.3. Chlorophyll a Fluorescence and Gas Exchange

The variables of fluorescence, potential quantum efficiency adapted to light (Fv'/Fm'), effective quantum efficiency ($\phi PSII = \Delta F/Fm'$) of photosystem II (PSII), electron transport ratio ($ETR = PPFD \times \Delta F/Fm' \times 0.5 \times 0.84$), heat dissipation in the antenna complex ($D = 1 - Fv'/Fm'$), and dissipation of excess energy from the PSII reaction center ($Ex = Fv'/Fm' \times (1 - qP)$) were determined [34,35]. The variables of gas exchange, CO_2 assimilation rate (*Anet*, μmol CO_2 $m^{-2}s^{-1}$), transpiration rate (*E*, mmol water vapor $m^{-2}s^{-1}$), stomatal conductance (*gs*, mmol $m^{-2}s^{-1}$), internal leaf CO_2 concentration (*Ci*, μmol m^{-2} s^{-1} Pa^{-1}), relative humidity (RH, %), instantaneous water-use efficiency (iWUE, μmol CO_2 (mmol H_2O)$^{-1}$), and RuBisCO carboxylation efficiency (*Anet/Ci* (μmol m^{-2} s^{-1} Pa^{-1}) [36] were determined using an open system of photosynthesis with a CO_2 analyzer and infrared water vapor radiation ("Infra-Red Gas Analyzer—IRGA", model GSF 3000, Walz, GmbH, Effeltrich, Germany) under a saturating irradiance of 1200 μmol m^{-2} s^{-1} (photosynthetic photon flux density (PPFD)), with a coupled modulated light fluorometer (LED-Array/PAM-Fluorometer 3055-FL, Walz, GmbH, Effeltrich, Germany). These measures were carried out in the period from 9:00 am to 11:00 am on a sunny day [37] with environmental CO_2 equal to 415 ± 10 μmol m^{-2} s^{-1} and the following light intensities in the dry and rainy seasons: Dry1: 874.3; Dry2: 484.24; Rainy1: 873.21; and Rainy2: 900.07 μmol m^{-2} s^{-1} PPFD.

2.4. *Water Potential and Relative Leaf Water Content*

The water potential, represented in Mpa, of fully expanded leaves of the same individual was determined using a WP4-T water potential analyzer with a temperature controller (Decagon Devices, Pullman, WA, USA).

The relative leaf water content (RWC = (FM − DM)/(TM − DM) × 100) was determined using the weighing method for the fresh mass (FM), turgid mass (TM), and dry mass (DM) [38,39].

2.5. *Yield and Chemical Profile of Essential Oil*

The leaves of *Xylopia aromatica* were separated and dried at room temperature; essential oils were extracted by means of hydro-distillation in a Clevenger apparatus for 2 h. The essential oil yield was calculated as the mass of essential oil (g) per mass of dry plant material (g) and later expressed as a percentage.

The chemical composition was determined using a gas chromatograph coupled to a mass spectrometer (GC-MS—Shimadzu, QP-5000, Kyoto, Japan) operating at 70 eV. It was equipped with a DB-5 fused silica capillary column (30 m × 0.25 mm × 0.25 μm), with helium as the carrier gas (1.0 mL·min^{-1}), an injector at 220 °C, a detector at 230 °C, a split of 1/20, and a temperature program of 60–240 °C, 3 °C min^{-1}.

The substances were identified by comparing the obtained GC-MS with the system database, the linear retention index (LRI), and data from the literature [34]. The linear retention index (LRI) values of the substances were obtained by analyzing of the mixture of n-alkanes (C9–C24-99%, Sigma Aldrich, São Paulo, Brazil) under the same operating conditions as the samples, applying the Van Den Dool and Kratz [35] equation.

The quantification of the essential oil was performed using gas chromatography with a flame ionization detector (GC-FID), with the area normalization method and the operating conditions described above.

2.6. *Photosynthetic Pigments, Carbohydrates, Antioxidant Enzymes, and Lipid Peroxidation*

Determinations were performed on fully expanded leaves. Photosynthetic pigments, chlorophyll a and b, carotenoids, and anthocyanin were extracted and quantified according to the methodology proposed in [40].

The reducing sugars [38], total soluble sugars [39,41], and starch [42] were also quantified.

The total soluble protein content [43]; the activities of peroxidase (POD, EC 1.11.1.7) [44], superoxide dismutase (SOD, EC 1.15.1.1), and catalase enzymes (CAT, EC 1.11.1.6) [45]; and lipid peroxidation [46] were also determined.

2.7. *Statistical Analysis*

Variables were evaluated via analysis of variance (ANOVA), and the means were compared using the Tukey test at 5% probability in the SigmaPlot software, version 12.0 [47].

3. Results

3.1. Water Relations, Chlorophyll a Fluorescence, and Gas Exchange

The water potential (Ψleaf) did not differ in the evaluated plants ($p > 0.741$). The relative humidity (RH%) and relative water content (RWC) were lower in the Dry1 season (Figure 1), while the vapor pressure deficit (VpdL) was greater (Figure 1). In general, in the rainy season (Rainy1 and Rainy2), the plants showed a higher electron transport ratio (ETR), effective quantum efficiency (ΦPSII), CO_2 assimilation rate (*Anet*), stomatal conductance (*gs*), RuBisCO carboxylation efficiency (*Anet/Ci*), and instantaneous water-use efficiency (iWUE). The dissipation of excess energy from the PSII reaction center (*Ex*) was higher in the dry season (August 2017) (Figures 2 and 3).

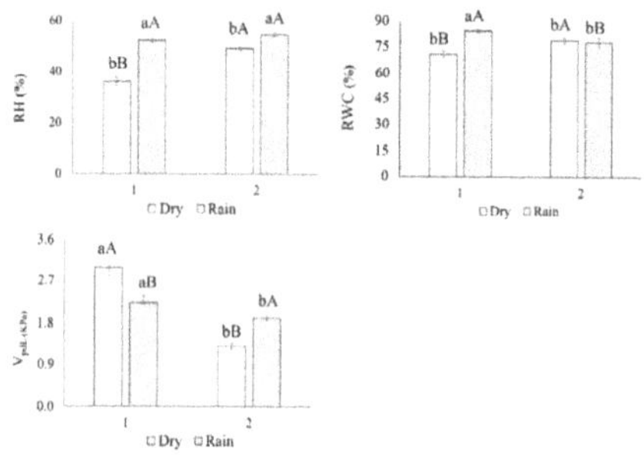

Figure 1. Relative humidity (RH, %) $p \leq 0.001$, relative water content (RWC, %) $p \leq 0.001$, and leaf vapor pressure deficit (VpdL, kPa) $p \leq 0.001$ of *Xylopia aromatica* evaluated in the dry season (Dry1 and Dry2) and in the rainy season (Rainy1 and Rainy2) in the Brazilian savanna of Botucatu, SP, Brazil. Medium values. Capital letters test data through the years, lowercase letters test dry and rainy seasons. Means followed by the same letter did not differ from each other in the Tukey test at 5% probability.

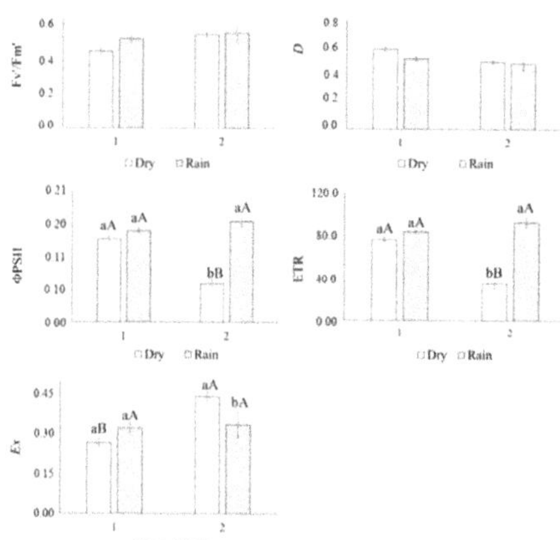

Figure 2. The efficiency of photosystem II (Fv′/Fm′) $p < 0.288$, the fraction of light absorbed by PSII antenna that is dissipated as heat (*D*) $p < 0.288$, effective quantum efficiency (ΦPSII) $p \leq 0.001$, electron transport rate (ETR) $p \leq 0.001$, and fraction of excitation energy not dissipated in the antenna that cannot be utilized for photochemistry (*Ex*) $p < 0.008$ of *Xylopia aromatica* evaluated in the dry season (Dry1 and Dry2) and in the rainy season (Rainy1 and Rainy2) in the Brazilian savanna of Botucatu, SP, Brazil. Medium values. Capital letters test data through the years, lowercase letters test dry and rainy seasons. Means followed by the same letter did not differ from each other in the Tukey test at 5% probability.

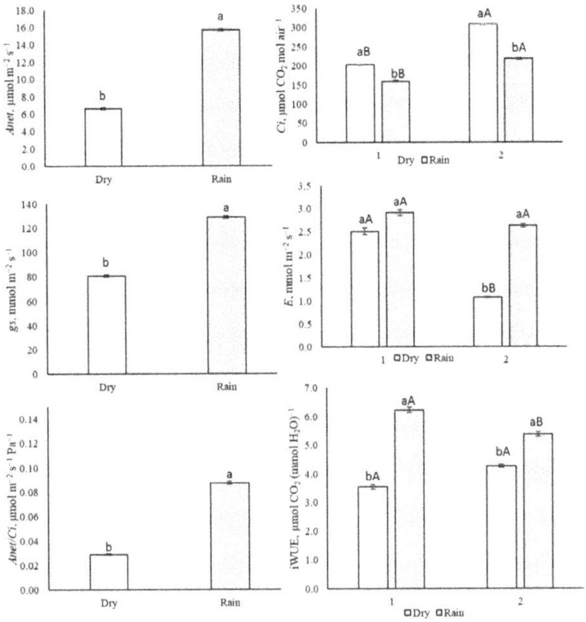

Figure 3. Leaf CO_2 assimilation rate (*Anet*, μmol m^{-2} s^{-1}) $p \leq 0.001$, accumulation in the substomatal chamber (*Ci*, μmol CO_2 mol Pa^{-1}) $p \leq 0.001$, stomatal conductance (*gs*, mmol m^{-2} s^{-1}) $p \leq 0.001$, transpiration rate (E, mmol m^{-2} s^{-1}) $p < 0.002$, carboxylation efficiency (*Anet/Ci*, μmol m^{-2} s^{-1} Pa^{-1}) $p \leq 0.001$, and instantaneous water-use efficiency (iWUE, μmol CO_2 (mmol H_2O^{-1}) $p < 0.005$ of *Xylopia aromatica* evaluated in the dry season (Dry1 and Dry2) and in the rainy season (Rainy1 and Rainy2) in the Brazilian savanna of Botucatu, SP, Brazil. Medium values. Capital letters test data through the years, lowercase letters test dry and rainy seasons. Means followed by the same letter did not differ from each other in the Tukey test at 5% probability.

3.2. Yield and Chemical Profile of Essential Oil

The yields of essential oil were 0.09% (Rainy1 season), 0.05% (Rainy2 season), 0.02% (Dry1 season), and 0.03% (Dry2 season), showing that the oil yield was greater in the rainy season ($p \leq 0.001$) (Figure 4).

Although the chemical profiles of the essential oils of the 28 individuals of *Xylopia aromatica* were similar, there was variation in the relative proportion of their constituents in the four evaluated groups for the dry season (Dry1 and Dry2) and rainy season (Rainy1 and Rainy2) (Supplementary Table S2 and Figure 4).

The chemical classes identified in the essential oils were monoterpene hydrocarbons (α-pinene, sabinene, β-pinene, myrcene, α-phellandrene, o-cymene, limonene, β-phellandrene, and cis-β-ocimene) with average concentrations as follows: Dry1 season (15.56%), Rainy1 season (7.29%), Dry2 season (0.22%), and Rainy2 season (1.45%); sesquiterpene hydrocarbons (α-copaene, cis-caryophyllene, aromadendrene, germacrene D, bicyclogermacrene, and α-muurolene) with average concentrations as follows: Dry1 season (28.07%), Rainy1 season (18.46%), Dry2 season (19.87%), and Rainy2 season (3.60%); and oxygenated sesquiterpenes (spathulenol, caryophyllene oxide, and globulol) with average concentrations as follows: Dry1 season (15.84%), Rainy1 season (30.06%), Dry2 season (41.9%), and Rainy2 season (61.09%) (Supplementary Table S2).

The major constituents of the essential oils were β-phellandrene, α-pinene and β-pinene, bicyclogermacrene, spathulenol, and an unidentified substance (sub7). The Dry1 season showed a higher relative proportion of β-phellandrene ($F = 7.157$ $p < 0.010$). The substances α-pinene (F = 4.289 $p < 0.043$) and β-pinene (F = 4.999 $p < 0.030$) were higher in the first evaluation of both the Dry1 and Rainy1 seasons (Figure 4).

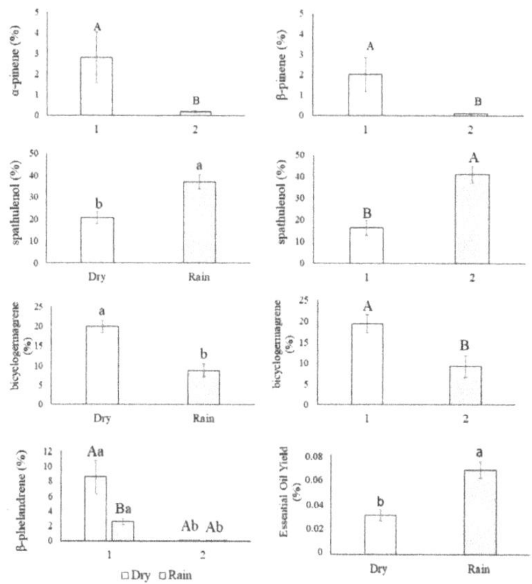

Figure 4. α-pinene ($p < 0.043$) and β-pinene ($p < 0.030$); spathulenol: season ($p \leq 0.001$), year ($p \leq 0.001$); bicyclogermacrene: season ($p \leq 0.001$), year ($p \leq 0.001$); β-phellandrene ($p < 0.012$), and essential oil yield of *Xylopia aromatica* evaluated in the dry season (Dry1 and Dry2) and in the rainy season (Rainy1 and Rainy2) in the Brazilian savanna of Botucatu, SP, Brazil. Medium values. Capital letters test data through the years, lowercase letters test dry and rainy seasons. Means followed by the same letter did not differ from each other in the Tukey test at 5% probability.

Bicyclogermacrene showed an increase in its relative percentage in the dry season and in evaluation 1, when the relative humidity was lower than in other evaluations (Supplementary Table S2 and Figure 4).

Spathulenol showed a higher relative proportion in the rainy season and assessment 2, a period of greater relative humidity (Supplementary Table S2 and Figure 4).

3.3. Photosynthetic Pigments, Carbohydrates, Antioxidant Enzymes, and Lipid Peroxidation

The chlorophyll a and b concentrations did not vary between seasons, while the carotenoid and anthocyanin concentrations were higher in the Dry2 season (Table 1).

Table 1. Photosynthetic pigments, chlorophyll *a* (μmol g^{-1}), chlorophyll *b* (μmol g^{-1}), carotenoids (μmol g^{-1}), and anthocyanins (μmol g^{-1}) in *Xylopia aromatica* (Lam.) Mart. evaluated in the dry season and in the rainy season, in the Cerrado of Botucatu, SP, southeastern Brazil.

Seasons	Chlorophyll *a*	Chlorophyll *b*	Carotenoids	Anthocyanins
Dry	676.45 A ± 30.93	307.67 A ±18.89	392.67 ± 14.76 A	427.81 ± 25.27 A
Rainy	737.14 A ± 18.30	296.06 A ± 17.61	355.43 ± 12.51 B	319.82 ±19.17 B
p	0.09	0.616	0.046	0.01

Averages followed by the same letter do not differ from each other. Tukey ($p < 0.05$). Mean data (n = 14).

The reducing sugars did not vary between seasons, and the concentrations of total sugars and starch were higher in the Dry2 season (Table 2).

Table 2. Total soluble sugar (mg of total sugar g^{-1} of FW), reducing sugar (mg of reducing sugar g^{-1} of FW), and starch (mg starch g^{-1} of FW) in *Xylopia aromatica* (Lam.) Mart. evaluated in the dry and rainy seasons in the Cerrado of Botucatu, SP, southeastern Brazil.

Seasons	Total Soluble Sugar	Reducing Sugar	Starch
Dry	4198.50 ± 132.96 A	4.987 A ± 0.214	107.95 ± 3.519 A
Rainy	3530.82 ± 192.46 B	5.344 A ± 0.172	69.84 ± 6.699 B
p	0.005	0.269	≤0.001

Averages followed by the same letter do not differ from each other. Tukey ($p < 0.05$). Mean data (n = 14).

The activities of the superoxide dismutase (SOD), peroxidase (POD), and catalase (CAT) enzymes and lipid peroxidation did not vary between seasons (Table 3).

Table 3. Superoxide dismutase (SOD, U mg^{-1} protein), peroxidase (POD, μmol of purpurogallin min^{-1} mg^{-1} protein), catalase (CAT, μKat μg^{-1} protein), and lipid peroxidation (MDA, malondialdehyde nmol g^{-1} FW) in *Xylopia aromatica* evaluated in the dry season and the rainy season in the Brazilian savanna of Botucatu, SP, southeastern Brazil.

Seasons	SOD	POD	CAT	MDA
Dry	38.84 A ± 2.945	0.061 A ± 0.00014	0.15 A ± 0.00031	65.63A ± 2.0274
Rainy	39.73 A ± 2.752	0.061 A ± 0.00057	0.13 A ±0.00018	64.63 A ± 2.475
p	0.785	0.937	0.515	0.936

Averages followed by the same letter do not differ from each other. Tukey ($p < 0.05$). Mean data (n = 14).

4. Discussion

The plants evaluated in the Rainy1 season showed a higher photochemical efficiency, indicative of high production of reducing agents, despite the large amount of energy not dissipated or used in the photochemical phase (*Ex*), which did not promote stress.

The CO_2 assimilation rate (Anet), stomatal conductance (gs), RuBisCO carboxylation efficiency (*Anet/Ci*), instantaneous water-use efficiency (iWUE), and relative water content (RWC) were higher in the Rainy1 season, when the plants presented lower vapor pressure deficits (VpdL). This condition allowed for greater stomatal apertures and increased the instantaneous water-use efficiency (iWUE) once the carbon input (*Anet,*) was greater than water output (E). These results agree with those recorded in [48].

In the Dry1 season, the lower relative humidity of the environment (RH%) and higher vapor pressure deficit (VpdL) resulted in partial closure of the stomata and lower stomatal conductance, a mechanism that contributed to the maintenance of water content and the regulation of leaf water potential (Ψleaf), which limited the entry of carbon and reduced its assimilation. These results agree with those obtained in [49] for Pinus palustres.

In the Dry2 season, plants showed the lowest effective quantum efficiency (ΦPSII) and electron transport rate (ETR), which may have resulted in lower production of reducing agents [50], limiting the synthesis of carbonic skeletons and causing the activation of alternative electron pathways, such as for the synthesis of carotenoids.

The greater amount of energy not dissipated or used in the photochemical phase (*Ex*) in the Dry2 season group should not be indicative of stress, since the concentrations of chlorophylls a and b did not change. The higher concentrations of carotenoids and anthocyanins must have helped the antioxidant enzymes in the neutralization of reactive oxygen species. Carotenoids in the xanthophyll pathway assist in the dissipation of photosystem II energy, which has the potential to cause oxidative stress [51].

These results agree with previous studies carried out in Eucalyptus globulus and Nicotiana attenuate [52,53]. In the present study, the increase in these pigments must have provided photoprotection and minimized the degradation of the membrane systems, since the amount of lipid peroxidation did not change.

Photoprotection can help prevent stress, which may explain the stability of antioxidant enzymes observed in the present study; this agrees with other results found in the literature [12,54].

Photoprotection can help prevent stress, which may explain the stability of the antioxidant enzymes observed in the present study; this agrees with other results found in the literature [12,54].

The finding of the highest concentration of starch in the dry season can be explained by the increased activity of the RuBisCO in the previous rainy season, storing surplus sugar for use in the unfavorable season, results that are confirmed in the literature [55]. This process probably maintains the concentration of reducing sugars in the dry and rainy seasons. These results suggest that in the rainy season, reducing sugars are derived from photosynthesis, while in the dry season, they come from the degradation of stored starch. These results suggest that the energy used in metabolism in the dry season may also be derived from starch degradation, as reported in [29].

In the dry season, *Xylopia aromatica* may direct part of its reducing sugars to the syntheses of carotenoids and anthocyanins, which would help explain the lower essential oil yield found in this study and in others [28,56,57].

Xylopia aromatica collected during the rainy season showed a higher electron transport rate, effective quantum efficiency, and stomatal conductance (*gs*), contributing to an increase in the rate of carbon assimilation and RuBisCO carboxylation efficiency (*Anet*/*Ci*). Therefore, the synthesis of a carbon skeleton for carbohydrate formation was possible, as revealed in a study carried out on rice [58]. It is suggested that, in the present study, carbohydrates formed in the rainy season provided carbon skeletons for essential oil synthesis. In this season the plants presented the highest effective quantum efficiency (ϕPSII), and instantaneous water-use efficiency (iWUE), which may have contributed to the higher leaf essential oil yield of *Xylopia aromatica* in the rainy season, as observed in [29].

In the Rainy1 season, a higher concentration of boron in the soil may have stimulated the synthesis of terpenes, evidenced by the greater production of essential oil. These results agree with those of [59], which suggested the participation of boron in moderate oxidative stress and essential oil production.

The relative proportions of the substances β-phellandrene, bicyclogermacrene, and spathulenol in the essential oil of *Xylopia aromatica* differed between sampling times, suggesting their influence on essential oil synthesis.

In the dry season, an unfavorable period for plant growth [60], there was a high proportion of β-phellandrene and bicyclogermacrene, suggesting the targeting of photosynthetic resources for the metabolism of these substances, which can help the activation of defense mechanisms and contribute to the acclimatization of the species to unfavorable conditions. These results verify previous findings [61,62].

Solanum lycopersicum presents β-phellandrene under osmotic stress conditions, and the substance is referred to as a stress indicator [31]. *Parthenium argentatum* has been shown to grow in conditions of moderate water deficit, displaying an increase in the relative percentage of β-phellandrene in its flower essential oil [32]. *Salvia dolomitica* cultivated in severe water deficit presented a two-fold increase in bicyclogermacrene content [33].

The high content of spathulenol in the rainy season is perhaps explained by the higher relative humidity, a favorable condition for the incidence and prevalence of phytopathogenic agents such as fungi [63]. Spathulenol presents antimicrobial activity [23,64], which can help control these pathogens.

5. Conclusions

Xylopia aromatica reveals photosynthetic and biochemical strategies that help it metabolically adjust to water availability, such as photoprotection mechanisms, stomatal control, starch accumulation, and terpene synthesis that contribute to the acclimatization of the species in the Brazilian savanna.

In the rainy season, better photosynthetic performance led to greater investment in essential oil production. In the dry season, the plants may direct part of their reducing sugars to the syntheses of carotenoids and anthocyanins, which must have helped the antioxidant enzymes neutralize reactive oxygen species. Carotenoids of the xanthophyll pathway assist in the dissipation of photosystem II energy, which can potentially cause oxidative stress. During this season, lower stomatal conductance prevented excessive water loss. These results suggest acclimatization of the species to the conditions of the Brazilian savanna.

Supplementary Materials: The following supporting information can be downloaded at: https://www.mdpi.com/article/10.3390/life14111416/s1, Figure S1: Chemical and physical characteristics of soil; Table S1: Chemical and physical characteristics of soil in the dry season—(september/2016—D1 and august/2017—D2) and in the rainy season—(february/2017—R1 and february/2018—R2), from the Rio Bonito region. Table S2: Relative percentage (%) of the essential oil substances of *Xylopia aromatica*.

Author Contributions: Conceptualization, F.C, M.V. and C.B.; methodology, F.C., M.V., M.S. and L.J.; validation, F.C., M.M. and C.B.; formal analysis, F.C., M.V. and M.S.; investigation, F.C., M.V., M.S. and L.J.; resources, G.F., M.M. and C.B.; writing—original draft preparation, F.C., M.V., M.S., G.F., M.M. and C.B.; writing—review and editing, F.C., M.V. and C.B.; supervision, C.B. All authors have read and agreed to the published version of the manuscript.

Funding: This study was financed in part by the Coordenação de Aperfeiçoamento de Pessoal de Nível Superior, Brazil—CAPES (Finance Code 001) and Pro-Rectory of Postgraduate Studies (PROPG-Unesp) to support scientific publication (Unesp—PROPG 033/2024).

Institutional Review Board Statement: Not applicable.

Informed Consent Statement: Not applicable.

Data Availability Statement: This published paper includes all data produced or analyzed during this project.

Conflicts of Interest: The authors declare no conflicts of interest.

References

1. Lopes dos Santos, G.; Pereira, M.G.; Delgado, R.C.; Magistrali, I.C.; Gomes da Silva, C.; Magno Moreira de Oliveira, C.; Pedro Bessa Larangeira, J.; Paula da Silva, T. Degradation of the Brazilian Cerrado: Interactions with Human Disturbance and Environmental Variables. *For. Ecol. Manag.* **2021**, *482*, 118875. [CrossRef]
2. Klink, C.A.; Machado, R.B. Conservation of the Brazilian Cerrado. *Conserv. Biol.* **2005**, *19*, 707–713. [CrossRef]
3. Dias, L.C.C.; Moschini, L.E.; Trevisan, D.P. A Influência das Atividades Antrópicas na Paisagem da Área de Proteção Ambiental Estadual do Rio Pandeiros, MG—Brasil. *Fronteiras* **2017**, *6*, 85–105. [CrossRef]
4. Beuchle, R.; Grecchi, R.C.; Shimabukuro, Y.E.; Seliger, R.; Eva, H.D.; Sano, E.; Achard, F. Land Cover Changes in the Brazilian Cerrado and Caatinga Biomes from 1990 to 2010 Based on a Systematic Remote Sensing Sampling Approach. *Appl. Geogr.* **2015**, *58*, 116–127. [CrossRef]
5. Fagundes, N.C.A.; Ferreira, E.J. Veredas Da Região Sudeste: Peculiaridades Florísticas e Estruturais e Situação de Conservação. *Neotrop. Biol. Conserv.* **2016**, *11*, 178–183. [CrossRef]
6. Maas, P.; Lobão, A.; Rainer, H. Annonaceae in Lista de Espécies da Flora do Brasil. 2015. Available online: http://floradobrasil2015.jbrj.gov.br/FB110557 (accessed on 1 November 2024).
7. Maia, J.G.S.; Andrade, E.H.A.; da Silva, A.C.M.; Oliveira, J.; Carreira, L.M.M.; Araújo, J.S. Leaf Volatile Oils from Four Brazilian *Xylopia* Species. *Flavour Fragr. J.* **2005**, *20*, 474–477. [CrossRef]
8. Junqueira, J.G.M.; Do Nascimento, M.N.G.; Da Costa, L.G.; Romualdo, L.L.; De Aquino, F.W.B.; Abubakar, M.N.; Terezan, A.P.; Cunha, G.O.S.; Severino, V.G.P. In Vivo and in Vitro Volatile Constituents of the Flowers of *Xylopia aromatica* by HS-SPME/GC-MS. *J. Braz. Chem. Soc.* **2021**, *32*, 1111–1119. [CrossRef]
9. Boni, T.S.; de Azevedo, J.P.; Oviedo Rodriguez, A.; Maltoni, K.L. *Xylopia aromatica*: Crescimento Inicial e Status Nutricional de Mudas Em Solo Degradado Condicionado Com Resíduos. *Res. Soc. Dev.* **2022**, *11*, e59211528582. [CrossRef]
10. Abbas, F.; Ke, Y.; Yu, R.; Yue, Y.; Amanullah, S.; Jahangir, M.M.; Fan, Y. Volatile Terpenoids: Multiple Functions, Biosynthesis, Modulation and Manipulation by Genetic Engineering. *Planta* **2017**, *246*, 803–816. [CrossRef]
11. Delazar, A.; Bahmani, M.; Shoar, H.H.; Tabatabaei-Raisi, A.; Asnaashari, S.; Nahar, L.; Sarker, S.D. Effect of Altitude, Temperature and Soil on Essential Oil Production in Thymus Fedtschenkoi Flowers in Osko and Surrounding Areas in Iran. *J. Essent. Oil-Bear. Plants* **2011**, *14*, 23–29. [CrossRef]

12. Spinelli, F.; Cellini, A.; Marchetti, L.; Mudigere, K.; Piovene, C. Emission and Function of Volatile Organic Compounds in Response to Abiotic Stress. In *Abiotic Stress in Plants—Mechanisms and Adaptations*; InTech Open: Rijeka, Croatia, 2011. [CrossRef]
13. Soran, M.L.; Stan, M.; Niinemets, Ü.; Copolovici, L. Influence of Microwave Frequency Electromagnetic Radiation on Terpene Emission and Content in Aromatic Plants. *J. Plant Physiol.* **2014**, *171*, 1436–1443. [CrossRef] [PubMed]
14. Fournier, G.; Hadjiakhoondi, A.; Charles, B.; Fourniat, J.; Leboeuf, M.; Cave, A. Chemical and Biological Studies of *Xylopia aromatica* Stem Bark and Leaf Oils. *Planta Med.* **1994**, *60*, 283–284. [CrossRef] [PubMed]
15. Da Silva, M.S.; Tavares, J.F.; Queiroga, K.F.; De Fátima Agra, M.; Barbosa Filho, J.M.; Da Silva Almeida, J.R.G.; Da Silva, S.A.S. Alcaloides e Outros Constituintes de *Xylopia langsdorffiana* (Annonaceae). *Quim. Nova* **2009**, *32*, 1566–1570. [CrossRef]
16. Bakkali, F.; Averbeck, S.; Averbeck, D.; Idaomar, M. Biological Effects of Essential Oils—A Review. *Food Chem. Toxicol.* **2008**, *46*, 446–475. [CrossRef]
17. Moura, A.P.G.; Beltrão, D.M.; Pita, J.C.L.R.; Xavier, A.L.; Brito, M.T.; de Sousa, T.K.G.; Batista, L.M.; de Carvalho, J.E.; Ruiz, A.L.T.G.; Della Torre, A.; et al. Essential Oil from Fruit of *Xylopia langsdorffiana*: Antitumour Activity and Toxicity. *Pharm. Biol.* **2016**, *54*, 3093–3102. [CrossRef]
18. Aati, H.; El-Gamal, A.; Kayser, O. Chemical Composition and Biological Activity of the Essential Oil from the Root of *Jatropha pelargoniifolia* Courb. Native to Saudi Arabia. *Saudi Pharm. J.* **2019**, *27*, 88–95. [CrossRef]
19. Oliveira, V.B.; Araújo, R.L.B.; Eidenberger, T.; Brandão, M.G.L. Chemical Composition and Inhibitory Activities on Dipeptidyl Peptidase IV and Pancreatic Lipase of Two Underutilized Species from the Brazilian Savannah: *Oxalis cordata* A.St.-Hil. and *Xylopia aromatica* (Lam.) Mart. *Food Res. Int.* **2018**, *105*, 989–995. [CrossRef]
20. Andrade, E.H.A.; da Silva, A.C.M.; Carreira, L.M.M.; Oliveira, J.; Maia, J.G.S. Essential Oil Composition from Leaf, Fruit and Flower of *Xylopia aromatica* (Lam.) Mart. *J. Essent. Oil Bear. Plants* **2004**, *7*, 151–154. [CrossRef]
21. do Nascimento, K.F.; Moreira, F.M.F.; Alencar Santos, J.; Kassuya, C.A.L.; Croda, J.H.R.; Cardoso, C.A.L.; do Carmo Vieira, M.; Góis Ruiz, A.L.T.; Ann Foglio, M.; de Carvalho, J.E.; et al. Antioxidant, Anti-Inflammatory, Antiproliferative and Antimycobacterial Activities of the Essential Oil of *Psidium guineense* Sw. and Spathulenol. *J. Ethnopharmacol.* **2018**, *210*, 351–358. [CrossRef]
22. Moreira Alcântara, J.; Mesquita, J.; De Lucena, V.M.; Facanali, R.; Ortiz, M.; Marques, M.; Da, M.; Lima, P. Chemical Composition and Bactericidal Activity of the Essential Oils of Four Species of *Annonaceae* Growing in Brazilian Amazon. *Nat. Prod. Commun.* **2017**, *12*, 619–622. [CrossRef]
23. Vieira, M.A.R.; Jorge, L.G.; Marçon, C.; Campos, F.G.; Rozada, A.M.F.; de Freitas Gauze, G.; Seixas, F.A.V.; Marques, M.O.M.; Mendes, R.P.; Boaro, C.S.F. Geographical Influences on the Chemical Composition and Antifungal Activity of *Xylopia aromatica* (Lam.) Mart. Leaf Essential Oil. *S. Afr. J. Bot.* **2023**, *160*, 209–218. [CrossRef]
24. Cao, B.-L.; Ma, Q.; Zhao, Q.; Wang, L.; Xu, K. Effects of Silicon on Absorbed Light Allocation, Antioxidant Enzymes and Ultrastructure of Chloroplasts in Tomato Leaves under Simulated Drought Stress. *Sci. Hortic.* **2015**, *194*, 53–62. [CrossRef]
25. Spychalla, J.P.; Desborough, S.L. Superoxide Dismutase, Catalase, and α-Tocopherol Content of Stored Potato Tubers. *Plant Physiol.* **1990**, *94*, 1214–1218. [CrossRef] [PubMed]
26. Erinle, K.O.; Jiang, Z.; Ma, B.; Li, J.; Chen, Y.; Ur-Rehman, K.; Shahla, A.; Zhang, Y. Exogenous Calcium Induces Tolerance to Atrazine Stress in Pennisetum Seedlings and Promotes Photosynthetic Activity, Antioxidant Enzymes and PsbA Gene Transcripts. *Ecotoxicol. Environ. Saf.* **2016**, *132*, 403–412. [CrossRef] [PubMed]
27. Apel, K.; Hirt, H. Reactive Oxygen Species: Metabolism, Oxidative Stress, and Signal Transduction. *Annu. Rev. Plant Biol.* **2004**, *55*, 373–399. [CrossRef]
28. Selmar, D.; Kleinwächter, M. Stress Enhances the Synthesis of Secondary Plant Products: The Impact of Stress-Related over-Reduction on the Accumulation of Natural Products. *Plant Cell Physiol.* **2013**, *54*, 817–826. [CrossRef]
29. Loreto, F.; Schnitzler, J.P. Abiotic Stresses and Induced BVOCs. *Trends Plant Sci.* **2010**, *15*, 154–166. [CrossRef]
30. Zhao, L.; Chang, W.; Xiao, Y.; Liu, H.; Liu, P. Methylerythritol Phosphate Pathway of Isoprenoid Biosynthesis. *Annu. Rev. Biochem.* **2013**, *82*, 497–530. [CrossRef]
31. Tomescu, D.; Şumălan, R.; Copolovici, L.; Copolovici, D. The Influence of Soil Salinity on Volatile Organic Compounds Emission and Photosynthetic Parameters of *Solanum lycopersicum* L. Varieties. *Open Life Sci.* **2017**, *12*, 135–142. [CrossRef]
32. Nik, Z.B.; Mirza, M.; Ghaffari, M. Effect of Drought Stress on Growth and Essential Oil Contents in *Parthenium argentatum* Gray. *J. Essent. Oil Bear. Plants* **2008**, *11*, 423–429. [CrossRef]
33. Caser, M.; Chitarra, W.; D'Angiolillo, F.; Perrone, I.; Demasi, S.; Lovisolo, C.; Pistelli, L.; Pistelli, L.; Scariot, V. Drought Stress Adaptation Modulates Plant Secondary Metabolite Production in *Salvia dolomitica* Codd. *Ind. Crops Prod.* **2019**, *129*, 85–96. [CrossRef]
34. Adams, R.P. *Identification of Essential Oil Components by Gas Chromatography*, 4th ed.; Allured Pub. Corp.: Carol Stream, IL, USA, 2017; ISBN 9781932633214.
35. van Den Dool, H.; Kratz, P.D. A Generalization of the Retention Index System Including Linear Temperature Programmed Gas—Liquid Partition Chromatography. *J. Chromatogr. A* **1963**, *11*, 463–471. [CrossRef] [PubMed]
36. Zhang, S.; Li, Q.; Ma, K.; Chen, L. Temperature-Dependent Gas Exchange and Stomatal/Non-Stomatal Limitation to CO_2 Assimilation of *Quercus liaotungensis* under Midday High Irradiance. *Photosynthetica* **2001**, *39*, 383–388. [CrossRef]
37. Elsheery, N.I.; Cao, K.-F. Gas Exchange, Chlorophyll Fluorescence, and Osmotic Adjustment in Two Mango Cultivars under Drought Stress. *Acta Physiol. Plant* **2008**, *30*, 769–777. [CrossRef]

38. Miller, G.L. Use of Dinitrosalicylic Acid Reagent for Determination of Reducing Sugar. *Anal. Chem.* **1959**, *31*, 426–428. [CrossRef]
39. Morris, D.L. Quantitative Determination of Carbohydrates with Dreywood's Anthrone Reagent. *Science (1979)* **1948**, *107*, 254–255. [CrossRef]
40. Sims, D.A.; Gamon, J.A. Relationships between Leaf Pigment Content and Spectral Reflectance across a Wide Range of Species, Leaf Structures and Developmental Stages. *Remote Sens. Environ.* **2002**, *81*, 337–354. [CrossRef]
41. Yemm, E.W.; Willis, A.J. The Estomation of Carboydrates in Plant Extracts by Anthrone. *Biochem. J.* **1954**, *57*, 508–514. [CrossRef]
42. Clegg, K.M. The Application of the Anthrone Reagent to the Estimation of Starch in Cereals. *J. Sci. Food Agric.* **1956**, *7*, 40–44. [CrossRef]
43. Bradford, M.M. A Rapid and Sensitive Method for the Quantitation of Microgram Quantities of Protein Utilizing the Principle of Protein-Dye Binding. *Anal. Biochem.* **1976**, *72*, 248–254. [CrossRef]
44. Teisseire, H.; Guy, V. Copper-Induced Changes in Antioxidant Enzymes Activities in Fronds of Duckweed (*Lemna minor*). *Plant Sci.* **2000**, *153*, 65–72. [CrossRef]
45. Peixoto, P.H.P.; Pimenta, D.S.; Cambraia, J. Alterações Morfológicas e Acúmulo de Compostos Fenólicos Em Plantas de Sorgo Sob Estresse de Alumínio. *Bragantia* **2007**, *66*, 17–25. [CrossRef]
46. Heath, R.L.; Packer, L. Photoperoxidation in Isolated Chloroplasts. *Arch. Biochem. Biophys.* **1968**, *125*, 189–198. [CrossRef] [PubMed]
47. Zar, J.H. *Biostatistical Analysis*, 5th ed.; Pearson Prentice Hall: Upper Saddle River, NJ, USA, 2010.
48. Sage, R.F.; Way, D.A.; Kubien, D.S. Rubisco, Rubisco Activase, and Global Climate Change. *J. Exp. Bot.* **2008**, *59*, 1581–1595. [CrossRef] [PubMed]
49. Addington, R.N.; Mitchell, R.J.; Oren, R.; Donovan, L. a Stomatal Sensitivity to Vapor Pressure Deficit and Its Relationship to Hydraulic Conductance in *Pinus palustris*. *Tree Physiol.* **2004**, *24*, 561–569. [CrossRef]
50. Sharma, A.; Kumar, V.; Shahzad, B.; Ramakrishnan, M.; Singh Sidhu, G.P.; Bali, A.S.; Handa, N.; Kapoor, D.; Yadav, P.; Khanna, K.; et al. Photosynthetic Response of Plants Under Different Abiotic Stresses: A Review. *J. Plant Growth Regul.* **2020**, *39*, 509–531. [CrossRef]
51. Baraldi, R.; Canaccini, F.; Cortes, S.; Magnani, F.; Rapparini, F.; Zamboni, A.; Raddi, S. Role of Xanthophyll Cycle-Mediated Photoprotection in *Arbutus unedo* Plants Exposed to Water Stress during the Mediterranean Summer. *Photosynthetica* **2008**, *46*, 378–386. [CrossRef]
52. Palmer-Young, E.C.; Veit, D.; Gershenzon, J.; Schuman, M.C. The Sesquiterpenes(*E*)-β-Farnesene and (*E*)-α-Bergamotene Quench Ozone but Fail to Protect the Wild Tobacco *Nicotiana attenuata* from Ozone, UVB, and Drought Stresses. *PLoS ONE* **2015**, *10*, e0127296. [CrossRef]
53. Machado, F.; Dias, M.C.; de Pinho, P.G.; Araújo, A.M.; Pinto, D.; Silva, A.; Correia, C.; Moutinho-Pereira, J.; Santos, C. Photosynthetic Performance and Volatile Organic Compounds Profile in Eucalyptus Globulus after UVB Radiation. *Environ. Exp. Bot.* **2017**, *140*, 141–149. [CrossRef]
54. Gill, S.S.; Tuteja, N. Reactive Oxygen Species and Antioxidant Machinery in Abiotic Stress Tolerance in Crop Plants. *Plant Physiol. Biochem.* **2010**, *48*, 909–930. [CrossRef]
55. Peng, Y.; Li, C.; Fritschi, F.B. Diurnal Dynamics of Maize Leaf Photosynthesis and Carbohydrate Concentrations in Response to Differential N Availability. *Environ. Exp. Bot.* **2014**, *99*, 18–27. [CrossRef]
56. Arimura, G.I.; Matsui, K.; Takabayashi, J. Chemical and Molecular Ecology of Herbivore-Induced Plant Volatiles: Proximate Factors and Their Ultimate Functions. *Plant Cell Physiol.* **2009**, *50*, 911–923. [CrossRef]
57. Chomel, M.; Guittonny-Larchevêque, M.; Fernandez, C.; Gallet, C.; DesRochers, A.; Paré, D.; Jackson, B.G.; Baldy, V. Plant Secondary Metabolites: A Key Driver of Litter Decomposition and Soil Nutrient Cycling. *J. Ecol.* **2016**, *104*, 1527–1541. [CrossRef]
58. Seneweera, S.; Makino, A.; Hirotsu, N.; Norton, R.; Suzuki, Y. New Insight into Photosynthetic Acclimation to Elevated CO_2: The Role of Leaf Nitrogen and Ribulose-1,5-Bisphosphate Carboxylase/Oxygenase Content in Rice Leaves. *Environ. Exp. Bot.* **2011**, *71*, 128–136. [CrossRef]
59. Choudhary, S.; Zehra, A.; Naeem, M.; Khan, M.M.A.; Aftab, T. Effects of Boron Toxicity on Growth, Oxidative Damage, Antioxidant Enzymes and Essential Oil Fingerprinting in Mentha Arvensis and *Cymbopogon flexuosus*. *Chem. Biol. Technol. Agric.* **2020**, *7*, 8. [CrossRef]
60. Llusia, J.; Roahtyn, S.; Yakir, D.; Rotenberg, E.; Seco, R.; Guenther, A.; Peñuelas, J. Photosynthesis, Stomatal Conductance and Terpene Emission Response to Water Availability in Dry and Mesic Mediterranean Forests. *Trees—Struct. Funct.* **2016**, *30*, 749–759. [CrossRef]
61. Llusià, J.; Peñuelas, J.; Alessio, G.A.; Estiarte, M. Seasonal Contrasting Changes of Foliar Concentrations of Terpenes and Other Volatile Organic Compound in Four Dominant Species of a Mediterranean Shrubland Submitted to a Field Experimental Drought and Warming. *Physiol. Plant* **2006**, *127*, 632–649. [CrossRef]
62. Dudareva, N.; Klempien, A.; Muhlemann, J.K.; Kaplan, I. Biosynthesis, Function and Metabolic Engineering of Plant Volatile Organic Compounds. *New Phytol.* **2013**, *198*, 16–32. [CrossRef]

63. Hussain, M.; Ismaili, N.J. Phytopathogenic fungi associated with ripening fruit of date palm (*Phoenix dactylifera* L.) during rainy season in the university area of Khairpur, Sindh, Pakistan. *Plant Prot.* **2019**, *3*, 161–165. [CrossRef]
64. Tan, N.; Satana, D.; Sen, B.; Altan, H.B.; Demirci, B.; Uzun, M. Antimycobacterial and Antifungal Activities of Selected Four Salvia Species. *Rec. Nat. Prod.* **2016**, *10*, 593–603.

life

MDPI

Article

Genetic Diversity and Population Structure Analysis of *Pinus elliottii* Germplasm Resources in Jiangxi Province

Min Yi, Rong Hu, Wending Huang, Tingxuan Chen, Wenlei Xie, Haiping Xie, Xin Luo and Meng Lai *

Jiangxi Provincial Key Laboratory of Subtropical Forest Resources Cultivation, 2011 Collaboration Innovation Center of Jiangxi Typical Trees Cultivation and Utilization, College of Forestry, Jiangxi Agricultural University, Nanchang 330045, China; yimin6104@163.com (M.Y.); hurong089@163.com (R.H.); hwd0208@163.com (W.H.); sea1067488960@163.com (T.C.); xwl981222@163.com (W.X.); 13302512932@163.com (H.X.); 18942335817@163.com (X.L.)

* Correspondence: laimeng21@163.com

Abstract: This study aimed to compare and assess the genetic diversity and trends among the introduced family provenance, first-cycle superior trees breeding provenance, and improved-generation superior trees breeding provenance of *Pinus elliottii* using EST-SSR markers. The goal was to provide a foundation for advanced genetic improvement and sustainable utilization of *P. elliottii* in Jiangxi Province. A total of 417 individuals were analyzed for their genetic diversity and population structure using 19 pairs of SSR markers. The analysis identified 103 alleles across all the samples, with an average of 5.421 alleles per locus. Compared to other coniferous species, *P. elliottii* exhibited a moderate to high level of genetic diversity ($I = 0.862$, $He = 0.457$). Analysis of the molecular variance (AMOVA) revealed that 97.90% of the genetic variation occurred within provenances, consistent with a low genetic differentiation coefficient ($Fst = 0.016 < 0.05$) and high gene flow ($Nm = 15.715$) among provenances. In addition, analysis using STRUCTURE v. 2.3.4 software divided the 417 germplasm samples into two distinct groups, corroborating the results of the principal coordinates analysis (PCoA) and the unweighted pair group method with arithmetic (UPGMA) clustering analysis. Overall, the germplasm resources of *P. elliottii* exhibited rich genetic diversity, with the majority of the genetic variation occurring within provenances. For the genetic improvement of high-resin-yielding slash pines, breeding programs should prioritize populations with high genetic diversity while carefully selecting superior individuals from within those populations. These findings provide a solid foundation for breeding high-resin-yielding varieties and for future research on the sustainable utilization of these valuable resources.

Keywords: *Pinus elliottii*; SSR markers; germplasm resources; genetic diversity; genetic differentiation; breeding strategies

1. Introduction

Pinus elliottii, native to the southeastern United States, is a valuable species for timber and pulp production, as well as a high-quality source of resin [1,2]. The resin from *P. elliottii* is renowned for its high yield, excellent quality, resistance to crystallization, slow coagulation, and low impurity content. Since its introduction into Jiangxi Province in 1948, *P. elliottii* has become a key afforestation species in the hilly areas of the Jitai Plain, primarily due to its strong adaptability, rapid growth, and high resin yield [3]. To accelerate the selection and breeding of high-resin-yielding *P. elliottii* varieties, genetic improvement efforts began in China in the 1980s. These initiatives have focused on genetic variation in resin-related traits, resin composition, and molecular breeding for resin production, laying the groundwork for advanced genetic improvement of *P. elliottii* [4–6]. Interspecific hybridization is a vital approach for germplasm innovation in pine trees, expanding genetic resources and incorporating favorable traits from both closely related and distantly related species. This strategy addresses the limitations of close-relative hybridization by broadening the gene pool [7]. Previous research has explored the cross-compatibility of *P. elliottii* with native Chinese pine species (*Pinus massoniana*) and exotic species (*Pinus taeda* and *Pinus caribaea*). The results indicated that three hybrid combinations—*Pinus caribaea* × *P. elliottii*, *P. elliottii* × *Pinus caribaea*, and *P. elliottii* × *Pinus taeda*—exhibited fertility levels equal to or higher than their corresponding half-sib controls. In contrast, the *Pinus massoniana* × *P. elliottii* and *Pinus taeda* × *P. elliottii* hybrids showed significantly lower fertility than the half-sib controls, with some exhibiting little to no fertility [8]. Furthermore, *P. elliottii* is typically characterized by dense foliage and large lateral branches during its juvenile phase, contributing to the severe damage during the 2008 snow disaster in southern China. Therefore, hybridization to improve the needle size and lateral branch traits holds significant potential for enhancing its resilience to such events [9,10].

However, due to the lack of natural *P. elliottii* germplasm resources in China, the genetic foundation for breeding improvements depends heavily on the diversity of the introduced germplasm. Early introductions were often driven by specific objectives, with limited consideration of population genetic diversity, which has limited the efficiency and effectiveness of *P. elliottii* breeding programs. As a result, a comprehensive study of the genetic base of breeding resources in Jiangxi Province is crucial to better understand the genetic information and background of the germplasm. This will facilitate the development of breeding populations with enhanced genetic quality and provide essential theoretical guidance for future genetic improvement and the creation of *P. elliottii* germplasm. Foundational studies on the genetic diversity of breeding populations have been conducted for many conifer species, both domestically and internationally, including Masson pine (*Pinus massoniana*) [11], Scots pine (*Pinus sylvestris*) [12], loblolly pine (*Pinus taeda*) [13], Korean pine (*Pinus koraiensis*) [14], and European larch (*Larix decidua*) [15].

Genetic diversity can be assessed through a variety of approaches, including morphological, cytological, physiological, biochemical, and molecular markers [16]. Early studies primarily used observable morphological traits to analyze genetic variation, focusing on phenotypic evaluation. However, morphological markers are limited in number, lack standardized observation criteria, and are highly influenced by environmental factors, making them less suitable for genetic variation studies [17]. In contrast, molecular markers represent the genetic diversity at the DNA level, offering greater precision as they are unaffected by environmental conditions [18]. Among the available molecular markers, simple sequence repeat (SSR) markers are particularly favored due to their abundance, high reliability, and convenient operation [19,20]. SSR markers have been widely applied in studies of genetic diversity, population structure, and phylogenetic relationships in forest tree germplasm, such as Masson pine [11], Scots pine [21], Chinese pine (*Pinus tabuliformis*) [22], and Korean pine [14]. Although SSR analysis of *P. elliottii* has been employed in studies such as transcriptomes [6,23], SSR marker development [24], genetic diversity studies using RAPD molecular markers in seed orchards [25], exome resequencing [26], and genetic map construction [27], comprehensive studies on the genetic diversity of *P. elliottii* germplasm resources remain limited.

Although our research group has collected a substantial amount of *P. elliottii* germplasm from both the United States and various regions across China, the genetic background of these diverse sources remains unclear. This has led to issues, such as germplasm admixture, genetic background ambiguity, and a lack of systematic integrated research, making it difficult to select suitable breeding materials. Therefore, assessing the genetic background of these populations is crucial. To ensure a robust genetic foundation for the sustainable improvement of *P. elliottii*, this study focused on the analysis of the genetic diversity and population structure across breeding provenances with different levels of improvement, using SSR molecular markers. The materials included the introduced family provenance, first-cycle superior trees breeding provenance, and improved-generation superior trees breeding provenance collected from Jiangxi Province. The aim was to elucidate the complexity of their genetic backgrounds, the relationships among materials, and the degree of genetic variation. This research provides a theoretical basis and material foundation for parent selection in hybrid breeding, as well as for constructing core germplasm and fingerprinting profiles. Moreover, it is expected to improve the efficiency of breeding high-resin-yielding *P. elliottii* varieties while laying the groundwork for further exploration and utilization of high-resin-yielding germplasm in the future.

2. Materials and Methods

2.1. Experimental Materials

The experimental materials for this study consisted of 417 *P. elliottii* germplasm resources currently preserved in Jiangxi Province. Samples were collected from young buds or leaves, which were immediately flash-frozen in liquid nitrogen and stored at −80 °C until further use. The primary sources of these samples are detailed below (Table 1).

Table 1. Information on the *P. elliottii* germplasm resources.

Provenance	Ways to Save	Source	Sample Number	Effective Amplification Number	Afforestation Site and Year
IP *	Introduced family trial forest	Georgia, Mississippi, and Florida, USA	113	112	Ji'an (1990)
		Florida, USA	30	8	Jingdezhen (2011)
		Florida, USA	10	10	Jingdezhen (2018)
		Kansas, USA	40	40	Ganzhou (2018)
FP	First-generation seed orchard	Early-introduced slash pine plantations in Jiangxi Province, China	44	44	Ji'an (1992)
			77	77	Xiajiang (1980)
IGP	Improved-generation seed orchard	Experimental forests planted with selected superior trees across Jiangxi province, China	47	47	Xiajiang (2015)
			79	79	Ji'an (2011)

* IP: Introduced family provenance; FP: First-cycle superior trees breeding provenance; IGP: Improved-generation superior trees breeding provenance, the same as below.

Introduced family provenance (IP): This group includes 113 families from the *P. elliottii* provenance/family trial forest established in 1990 at Baiyunshan Forest Farm (27.22° N, 115.13° E), Qingyuan District, Ji'an City. The seeds were sourced from seed orchards in Georgia, Mississippi, and Florida, USA, with a total of 112 samples successfully amplified. Additionally, a family trial forest established in 2011 and 2018 at Fengshushan Forest Farm (29.37° N, 117.25° E), Jingdezhen, includes 30 and 10 families, respectively, also sourced from seed orchards in Florida, USA, with 18 samples successfully amplified. Furthermore, the superior family trial forest established in 2018 by the Ganzhou Forestry Research Institute (25.38° N, 114.93° E) includes 40 families sourced from seed orchards in Arkansas, USA, with all 40 samples successfully amplified.

First-cycle superior trees breeding provenance (FP): This group primarily consists of slash pine plantation stands from early introductions in Jiangxi Province. A total of 44 samples were collected from the first-generation seed orchard established in May 1992 at Baiyunshan Forest Farm, Qingyuan District, Ji'an City. An additional 77 samples were collected from the first-generation seed orchard established in 1980 at Xiajiang County Superior Tree Breeding Farm (27.58° N, 115.32° E).

Improved-generation superior trees breeding provenance (IGP): This group includes 126 samples from experimental forests planted with selected superior trees from across the province, with 79 samples from Baiyunshan Forest Farm, Qingyuan District, Ji'an City, and 47 samples from Xiajiang County Superior Tree Breeding Farm.

2.2. Research Methods

2.2.1. Genomic DNA Extraction and Detection

Genomic DNA was extracted using the DNAsecure Plant Kit (DP320) from Tiangen Biotech Co., Ltd. (Beijing, China). The concentration and quality of the extracted DNA were assessed via 1% agarose gel electrophoresis. The DNA samples were then diluted to a uniform concentration of 20–50 ng/μL and stored at −20 °C for further analysis.

2.2.2. SSR Primer Screening

A total of 24 pairs of EST-SSR primers were selected from the published literature [24]. Additionally, SSR loci were identified within unigenes obtained from the *P. elliottii* transcriptome using MISA software (http://misaweb.ipk-gatersleben.de/, accessed on 25 October 2024), with the following criteria: dinucleotide repeats > 6 times, and trinucleotide, tetranucleotide, pentanucleotide, and hexanucleotide repeats > 5 times. Based on the conserved regions flanking the SSR loci, 103 pairs of SSR primers were designed using Primer 3.0 software [28] and synthesized by Sangon Biotech Co., Ltd. (Shanghai, China). The primers were validated for polymorphism using 11 slash pine samples.

2.2.3. PCR Amplification and Capillary Electrophoresis

The PCR reaction system and program are detailed in Table 2. Primers that produced clear and polymorphic amplification bands were labeled with one of four fluorescent dyes (FAM, Rox, Tem, or Hex). The PCR products were then analyzed using capillary electrophoresis on an ABI 3730XL DNA sequencer in collaboration with Sangon Biotech Co., Ltd. (Shanghai, China).

Table 2. Reaction system and SSR-PCR procedure for *P. elliottii*.

Reaction System		Reaction Procedure		
Component and Concentration	**Volume (μL)**	**Cycling Parameters**	**Time (min, s)**	
10 × buffer (plus Mg^{2+})	2	94 °C initial denaturation	5 min	
10 mol/L dNTPs	1.6	94 °C denaturation	30 s	30 cycles
forward primer	0.8	56–60 °C annealing	30 s	30 cycles
reverse primer	0.8	72 °C extension	60 s	30 cycles
Taq DNA polymerase (5 μ/μL)	0.1	72 °C final extension	7 min	
ddH_2O	13.7	4 °C keep		
Template DNA (50 ng/μL)	1			

2.2.4. Statistical Analysis

Microsatellite allele data for each *P. elliottii* sample, obtained from capillary electrophoresis, were analyzed using GeneMarker v.2.2.0 software [29]. DataFormater v.2.7 [30] and GenAlEx v.6.5 software [31] were then used to convert the resulting data into the necessary formats for further analysis.

The genetic diversity parameters, including the number of alleles (*Na*), the effective number of alleles (*Ne*), Shannon's information index (*I*), observed (*Ho*) and expected heterozygosity (*He*), F-statistics (*Fis, Fit* and *Fst*) and gene flow (*Nm*), were calculated using POPGENE v.1.3.1 software [32]. The allelic frequency variation and polymorphic information content (PIC) were assessed and calculated using PowerMarker v.3.25 software [33].

To assess the degree of genetic variation within and between the three *P. elliottii* provenances, an analysis of molecular variance (AMOVA) was performed using ARLEQUIN v.3.5.2.2 software [34].

The genetic structure of the 417 individuals was analyzed using STRUCTURE v.2.3.4 software [35]. The number of clusters (K) was set from 1 to 10, with a burn-in period of 100,000 iterations and 10,000 MCMC repetitions. Each K value was run three times, and the optimal number of clusters was determined by plotting K against ΔK and selecting the peak value.

The genetic distances between samples or provenances were calculated using PowerMarker v.3.25 software, and a UPGMA dendrogram was constructed based on these distances using MEGA v.11.0 software [36]. In addition, principal coordinate analysis (PCoA) was conducted using GenAlEx v.6.5 software to further explore the genetic relationships between provenances based on the microsatellite allele data.

3. Results

3.1. Polymorphism Analysis of EST-SSR Markers

The capillary electrophoresis results showed that out of the 127 primer pairs tested, 112 successfully amplified bands, yielding an amplification success rate of 88.19%. Among these, 19 primer pairs were identified as highly polymorphic SSR markers, with a polymorphic loci ratio of 14.96% (Figure 1; Table 3).

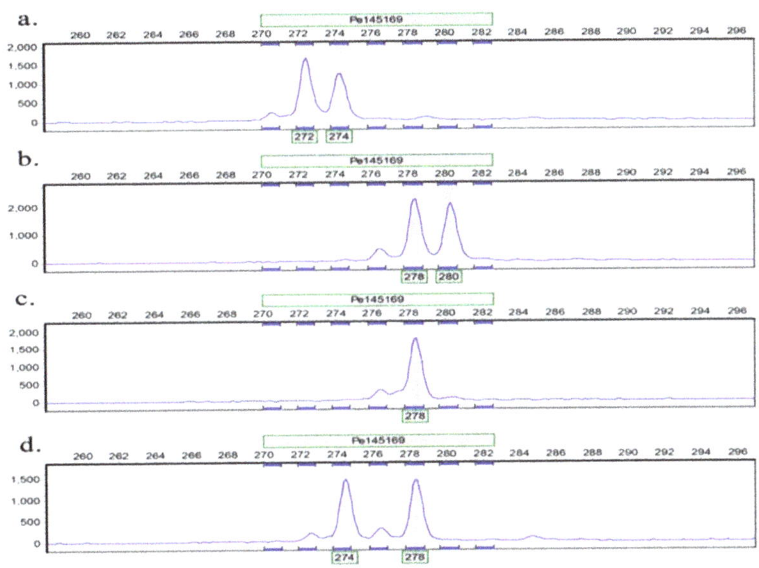

Figure 1. Amplification of some samples on primer Pe145169. The horizontal axis represents the size of the product (bp); The vertical axis represents the relative fluorescence signal; (**a**–**d**) represent four different samples.

Table 3. Sequence information of 19 pairs of polymorphic EST-SSR primers.

Locus	Repeat Motif	Product Size (bp)	Primer Sequence (5′-3′)
Pe145169	(CT)11	257	F: TGAATCCTCGGAATTTCTGG R: TGCCATTGAAACAAGCTGAA
Pe144426	(AT)8	152	F: AATGCGAGTGGCAACAAAGT R: ATTTCACATTCCCGTTCTCG
Pe138370	(CCCTG)5	179	F: CCTGACGCAACATAATCCCT R: ATAAAAGACACACCCCGCAG
Pe130187	(TC)6	213	F: TTCTCATGCTAAGCACACGC R: ATTTCTTCCATGGGTTCGTG
Pe132622	(AC)10	262	F: TATAACGTCAGCCCAGGGAC R: TTGCTTCTGCAGGAAAGGTT

Table 3. *Cont.*

Locus	Repeat Motif	Product Size (bp)	Primer Sequence (5′-3′)
Pe146453	(CATT)5	257	F: CTGATCCACCCTCATCTGCT R: GGGAGCAACCAGAACAACAT
Pe134815	(CTG)7	181	F: CCCCAAACCCCAACTTAGAT R: AAGTGGGAAAAATGAGGGCT
Pe131259	(AT)7	183	F: GGATTGATCCAAGCCAACAG R: ACCCGGAGGCAAATCTATCT
Pe103802	(AT)7	141	F: GGATGATCAGGGCATGAAAT R: CATAAAAGTTGGCACCACCA
Pe123077	(AAT)6	129	F: ATTGGGTTGAATCCGAACAT R: CCAGACAAAATTGTGGCCTT
Pe106732	(AGG)5	163	F: CGGTGGAAGATTTAGGTCCA R: GAAAAACAGCGGCAGAAAAG
Pe145380	(AC)9	244	F: CAACATTTGCTGTGAGCGTT R: ATGCATCCCTGATGCTCTTC
Pe119033	(AT)6	267	F: TTCTTGATACATCGGGGCAT R: AAACCTGTTCAAATCCTCACAA
Pe135178	(TTC)5	145	F: ATTTCAGAAGGTCAATGCGG R: GCAGGACATAAATGGGCAGT
Pe140688	(GATG)5	264	F: ATGAACGCTTTAGTTCCCCC R: GTGATGCGAGATGTGCAGTT
Pe110222	(TTG)5	277	F: TCTGTAACTTGGACTGGCCC R: CAGCCACAGTAGGTGCAACA
Pe113019	(AG)7	240	F: ATCTAGCGATCCCGGAAGTT R: ACCACCTTCTTCCTCCCATT
Pe139538	(TTAG)5	280	F: TGAAAGGTGGAGATCCTTGG R: AGGTCTGAGAGCATGGAGGA
Pe114582	(TA)6	275	F: ATACCTAGGCAGATGCCCCT R: TTAGGCTGGACAACCCAAAC

3.2. Cross-Species Applicability of EST-SSR Markers

To evaluate the cross-species applicability of the 19 polymorphic SSR markers developed for *P. elliottii*, 15 samples each of *Pinus massoniana* and *Pinus taeda* were selected for validation. The results (Table 4) demonstrated that all 19 SSR markers successfully amplified bands in both species, achieving a cross-species amplification rate of 100%.

Table 4. Transferability and percentage of the polymorphic loci for the tested SSR markers.

Sample	Amplification and Polymorphic (pair)	Amplification but No Polymorphism (pair)	No Amplification (pair)	Applicability (%)	Proportion of Polymorphic Sites (%)
Pinus massoniana	19	0	0	100%	100%
Pinus taeda	17	2	0	100%	89.47%

3.3. Genetic Diversity Analysis of Loci

A total of 103 alleles were detected across the 417 *P. elliottii* samples using the 19 pairs of SSR primers, with an average of 5.421 alleles per locus (Table 5). Primer Pe103802 exhibited the highest number of alleles (13), with an effective number of alleles of 1.777, followed by primer Pe113019, which had 8 alleles and an effective allele number of 1.263. Primers Pe123077 and Pe139538 had the fewest alleles, with two each, and effective allele numbers of 1.964 and 1.352, respectively. Shannon's information index (*I*) ranged from 0.370 (Pe114582) to 1.638 (Pe145169), with an average value of 0.862. The observed heterozygosity (*Ho*) ranged from 0.138 (Pe114582) to 0.795 (Pe123077), while the expected heterozygosity (*He*) ranged from 0.149 (Pe114582) to 0.777 (Pe145169), with average values of 0.402 and 0.457, respectively. For most loci, *He* was greater than *Ho*, except for primers

Pe119033, Pe106732, and Pe123077, indicating varying degrees of heterozygote deficiency. The polymorphism information content (PIC) values ranged from 0.146 (Pe114582) to 0.743 (Pe145169), with an average of 0.411. Among the loci, 3 had PIC values below 0.25, indicating low polymorphism; 10 loci had PIC values between 0.25 and 0.5, indicating moderate polymorphism; and 6 loci had PIC values greater than 0.5, indicating high polymorphism. Overall, these 19 SSR markers proved highly effective for analyzing the genetic diversity of the *P. elliottii* provenances.

Table 5. Diversity parameters of 19 microsatellite loci in *P. elliottii*.

Locus	*Na* *	*Ne*	*I*	*Ho*	*He*	PIC
Pe145169	7	4.491	1.638	0.636	0.777	0.743
Pe144426	7	3.361	1.332	0.448	0.702	0.651
Pe138370	5	2.800	1.168	0.564	0.643	0.580
Pe130187	7	2.386	1.200	0.505	0.581	0.538
Pe132622	7	2.275	1.068	0.498	0.560	0.511
Pe146453	4	2.461	1.006	0.352	0.594	0.513
Pe134815	7	1.879	0.947	0.451	0.468	0.436
Pe131259	5	1.956	0.908	0.441	0.489	0.444
Pe103802	13	1.777	0.978	0.408	0.437	0.413
Pe123077	2	1.964	0.684	0.795	0.491	0.370
Pe106732	4	1.724	0.836	0.430	0.420	0.394
Pe145380	5	1.799	0.789	0.383	0.444	0.385
Pe119033	6	1.660	0.836	0.423	0.397	0.373
Pe135178	3	1.688	0.605	0.368	0.408	0.325
Pe140688	3	1.512	0.578	0.224	0.339	0.293
Pe110222	3	1.450	0.571	0.264	0.311	0.281
Pe113019	8	1.263	0.445	0.159	0.208	0.194
Pe139538	2	1.352	0.429	0.144	0.260	0.226
Pe114582	5	1.175	0.370	0.138	0.149	0.146
Mean	5.421	2.051	0.862	0.402	0.457	0.411

*: *Na*: number of alleles; *Ne*: number of effective alleles; *Ho*: observed heterozygosity; *He*: expected heterozygosity; *I*: Shannon's information index; PIC: polymorphism information content, the same below.

3.4. Comparison of Genetic Diversity Among Provenances

A total of 83 and 95 alleles were detected in the FP and IGP of *P. elliottii*. In contrast, only 82 alleles were found in the introduced family provenance (Table 6). The ratios of the effective number of alleles (*Ne*) to the total number of alleles (*Na*) were higher in both the FP and IGP compared to the IP. Notably, the IP had a higher number of alleles at locus 135178 than the other two provenances. The FP exhibited the highest number of alleles at loci 144426 and 131259, while the IGP showed the highest number of alleles at seven loci: 145380, 119033, 114582, 138370, 134815, 113019, and 103802.

The Shannon's information index (*I*) values for the IP, FP, and IGP were 0.734, 0.839, and 0.964, respectively. The PIC values were 0.356, 0.407, and 0.460, respectively. The average *Ho* values were 0.371, 0.421, and 0.424, while the average *He* values were 0.398, 0.457, and 0.512, respectively. Overall, the genetic diversity of the FP and IGP was higher than that of the IP. These findings suggest that the *P. elliottii* provenances have maintained a relatively high level of genetic diversity following selective breeding and improvement efforts.

Table 6. Polymorphism analysis of 19 pairs of primers in three provenances.

Locus	Provenance	*Na*	*Ne*	*I*	*Ho*	*He*	PIC
Pe146453	IP	4	2.475	0.997	0.357	0.596	0.519
	FP	4	2.195	0.866	0.422	0.544	0.440
	IGP	4	2.575	1.078	0.277	0.612	0.545
Pe145380	IP	3	1.647	0.657	0.389	0.393	0.335
	FP	4	1.844	0.775	0.314	0.458	0.390
	IGP	5	1.973	0.907	0.447	0.493	0.437
Pe145169	IP	7	4.273	1.565	0.645	0.766	0.728
	FP	7	3.858	1.475	0.708	0.741	0.695
	IGP	7	5.125	1.765	0.553	0.805	0.779
Pe144426	IP	5	3.270	1.271	0.518	0.694	0.637
	FP	6	3.189	1.312	0.298	0.686	0.635
	IGP	5	3.518	1.373	0.500	0.716	0.668
Pe140688	IP	2	1.302	0.394	0.161	0.232	0.205
	FP	3	1.514	0.625	0.314	0.340	0.309
	IGP	3	1.780	0.662	0.222	0.438	0.349
Pe132622	IP	5	2.158	0.968	0.453	0.537	0.484
	FP	6	2.130	1.006	0.513	0.531	0.479
	IGP	6	2.573	1.181	0.544	0.611	0.564
Pe131259	IP	4	1.781	0.787	0.455	0.438	0.395
	FP	5	1.928	0.903	0.525	0.481	0.432
	IGP	4	2.177	0.974	0.344	0.541	0.488
Pe130187	IP	7	1.998	1.022	0.518	0.500	0.467
	FP	7	2.606	1.259	0.525	0.616	0.572
	IGP	7	2.646	1.259	0.468	0.622	0.566
Pe123077	IP	2	1.867	0.657	0.708	0.464	0.357
	FP	2	1.976	0.687	0.891	0.494	0.372
	IGP	2	2.00	0.693	0.815	0.500	0.375
Pe119033	IP	5	1.444	0.588	0.343	0.308	0.279
	FP	5	1.610	0.794	0.376	0.379	0.359
	IGP	6	2.045	1.048	0.571	0.511	0.480
Pe114582	IP	4	1.116	0.261	0.108	0.104	0.102
	FP	4	1.193	0.363	0.174	0.162	0.155
	IGP	5	1.241	0.467	0.142	0.194	0.189
Pe139538	IP	2	1.079	0.162	0.076	0.074	0.071
	FP	2	1.371	0.442	0.223	0.270	0.234
	IGP	2	1.734	0.614	0.160	0.423	0.334
Pe138370	IP	4	2.501	1.051	0.575	0.600	0.526
	FP	4	3.263	1.268	0.583	0.694	0.638
	IGP	5	2.736	1.165	0.532	0.634	0.572
Pe135178	IP	3	1.611	0.582	0.327	0.379	0.310
	FP	2	1.642	0.580	0.400	0.391	0.315
	IGP	2	1.823	0.644	0.393	0.451	0.350
Pe134815	IP	6	1.804	0.872	0.464	0.446	0.408
	FP	4	1.790	0.853	0.455	0.441	0.409
	IGP	7	2.056	1.060	0.432	0.514	0.481
Pe113019	IP	4	1.160	0.297	0.124	0.138	0.131
	FP	3	1.320	0.425	0.116	0.243	0.215
	IGP	8	1.356	0.594	0.246	0.262	0.248
Pe110222	IP	2	1.209	0.315	0.117	0.173	0.158
	FP	3	1.751	0.726	0.451	0.429	0.370
	IGP	3	1.533	0.647	0.286	0.348	0.319
Pe106732	IP	4	1.565	0.683	0.391	0.361	0.326
	FP	4	1.403	0.596	0.275	0.287	0.271
	IGP	4	2.351	1.092	0.632	0.575	0.535
Pe103802	IP	9	1.545	0.809	0.329	0.353	0.338
	FP	8	1.973	0.979	0.438	0.493	0.449
	IGP	10	1.918	1.083	0.484	0.479	0.456
Mean	IP	4.316	1.885	0.734	0.371	0.398	0.356
	FP	4.368	2.029	0.839	0.421	0.457	0.407
	IGP	5.000	2.272	0.964	0.424	0.512	0.460
Total	-	4.561	2.062	0.845	0.405	0.456	0.408

3.5. Genetic Differentiation and Gene Flow Among Provenances

As shown in Table 7, the *Fis* values across the 19 SSR loci ranged from −0.655 (Pe123077) to −0.401 (Pe139538), with an average of 0.110, indicating the presence of heterozygote deficiency within the provenances. The average *Ho* (0.402) was lower than the average *He* (0.457), further confirming the occurrence of heterozygote deficiency. The *Fit* values ranged from −0.634 (Pe123077) to −0.451 (Pe139538), with an average of 0.124. The overall *Fst* values ranged from 0.005 (Pe145380) to 0.084 (Pe139538), with an average of 0.016, indicating low genetic differentiation among the *P. elliottii* provenances. The gene flow (*Nm*) values ranged from 2.710 (Pe139538) to 53.260 (Pe145380), with an average of 15.715, suggesting substantial gene flow between the provenances, which likely reduces the genetic drift and contributes to the observed low genetic differentiation.

Table 7. Genetic differentiation and gene flow in the provenances of *P. elliottii*.

Locus	*Fis* *	*Fit*	*Fst*	*Nm*	Locus	*Fis*	*Fit*	*Fst*	*Nm*
Pe145169	0.175	0.183	0.009	26.595	106732	−0.061	−0.019	0.039	6.123
Pe144426	0.373	0.376	0.006	43.219	145380	0.144	0.148	0.005	53.260
Pe138370	0.124	0.131	0.008	30.356	119033	−0.078	−0.060	0.017	14.541
Pe130187	0.131	0.145	0.017	14.397	135178	0.083	0.090	0.008	30.815
Pe132622	0.101	0.106	0.006	45.202	140688	0.310	0.335	0.037	6.550
Pe146453	0.397	0.406	0.015	16.605	110222	0.100	0.129	0.032	7.517
Pe134815	0.036	0.042	0.006	40.009	113019	0.244	0.252	0.010	25.888
Pe131259	0.094	0.107	0.014	17.531	139538	0.401	0.451	0.084	2.710
Pe103802	0.055	0.067	0.012	20.022	114582	0.079	0.086	0.007	34.196
Pe123077	−0.655	−0.634	0.013	19.125	Mean	0.110	0.124	0.016	15.715

* *Fis*: inbreeding coefficient at the population level; *Fit*: inbreeding coefficient at total populations; *Fst*: proportion of differentiation among populations, the same below.

The AMOVA analysis based on the 19 loci (Table 8) revealed that the genetic variation among the provenances accounted for only 2.10% of the total variation, while 97.90% of the genetic variation was attributable to differences within individuals and among individuals within provenances. This indicated that most of the genetic variation in *P. elliottii* occurred within individuals, rather than between the IP, FP, and IGP.

Table 8. Analysis of the molecular variance (AMOVA) of *P. elliottii*.

Source	Degrees of Freedom	Sum of Squares	Variance Components	Percent of Variation (%)
Among provenance	2	54.576	0.085	2.10%
Within provenance	831	3289.002	3.958	97.90%
Total	833	3343.578	4.043	100%

The genetic similarity coefficients and genetic distances among the three provenances were calculated using Nei's method (Table 9). The genetic distance between the IP and the FP was 0.014, with a genetic similarity coefficient of 0.986. The genetic distance between the IP and the IGP was 0.020, with a genetic similarity coefficient of 0.980. The genetic distance between the FP and the IGP was 0.019, with a genetic similarity coefficient of 0.981. These results indicate that the IP is more closely related to the FP than to the IGP.

Table 9. Genetic distance (lowerleft) and genetic similarity (upperright) between three provenances.

Provenance	IP	FP	IGP
IP	****	0.986	0.980
FP	0.014	****	0.981
IGP	0.020	0.019	****

3.6. Population Genetic Structure Analysis

To further investigate the population genetic structure and relationships among the 417 *P. elliottii* samples, STRUCTURE software was used to analyze the data based on 19 pairs of SSR primers. The results (Figure 2) showed that as the K value increased, the lnP(D) also increased continuously. The most significant change in the lnP(D) occurred when the K value increased from 1 to 2, with the ΔK reaching its maximum peak at K = 2, indicating that the 417 *P. elliottii* samples were divided into two distinct subgroups (Figure 3).

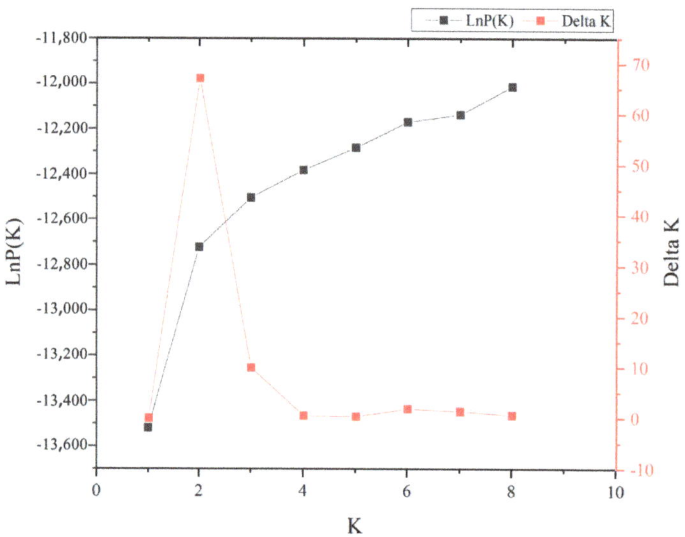

Figure 2. Diagnostic plots of the LnP(D) and ΔK from the STRUCTURE analysis of the *P. elliottii* provenances.

Figure 3. Results of the structure analysis of the *P. elliottii* provenances when K = 2. Red represents Subgroup I, blue represents Subgroup II.

To better characterize the genetic background of the materials, a Q value > 0.6 was used as the threshold for assigning the 417 *P. elliottii* samples into different subgroups. Based on this criterion, 410 samples were assigned to either Subgroup I or Subgroup II, indicating the relatively homogeneous genetic backgrounds among these samples. The 417 samples originated from the IP, FP and IGP. However, they were not strictly grouped according to their original provenances. Thus, the origins of the materials within each subgroup were further analyzed (Table 10). Subgroup I comprised 84.41% of the total samples, making it the most abundant group, with 352 samples derived from the introduced family trial forest (159 samples), the first-generation seed orchard (105 samples), and the advanced-generation seed orchard (88 samples). Subgroup II included 58 samples, with 4 from the introduced family trial forest, 16 from the first-generation seed orchard, and 38 from the advanced-generation seed orchard. Additionally, seven samples from the introduced family trial forest were placed in a mixed group, as they displayed mixed ancestry and complex genetic structures. These findings suggest that caution should be exercised when selecting these samples as breeding parents.

Table 10. Statistics of the population structure for 417 *P. elliottii*.

Provenance	Ways to Save	Subgroup I	Subgroup II	Mix Group
IP	Introduced family trial forest	159	4	7
FP	First-generation seed orchard	105	16	0
IGP	Advanced-generation seed orchard	88	38	0

3.7. PCoA and Genetic Clustering

To further analyze the cluster patterns, PCoA of *P. elliottii* was performed based on the pairwise genetic distance matrix of 19 EST-SSRs, with the results shown in Figure 4. The analysis revealed that the first principal coordinate (Coord. 1) and second principal coordinate (Coord. 2) accounted for 11.51% and 17.69% of the genetic variation, respectively. The individuals from the IP, FP, and IGP could not be completely separated, suggesting relatively close genetic relationships among these provenances. The entire *P. elliottii* germplasm could be roughly divided into two groups, which is consistent with the results of the STRUCTURE analysis at K = 2.

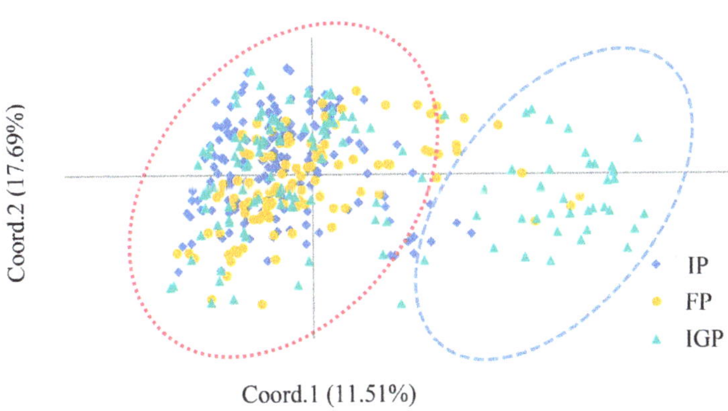

Figure 4. Principal coordinates analysis based on Nei's genetic distance of *P. elliottii*.

A clustering dendrogram of the 417 *P. elliottii* germplasm samples was constructed using the unweighted pair group method with arithmetic mean (UPGMA) in MEGA software. As indicated by the clustering results (Figure 5), the *P. elliottii* germplasm did not cluster strictly according to the IP, FP, and IGP. Instead, the samples were broadly divided into two groups. Except for a small number of samples with discrepancies, the clustering results were largely consistent with the findings from the PCoA and STRUCTURE analyses.

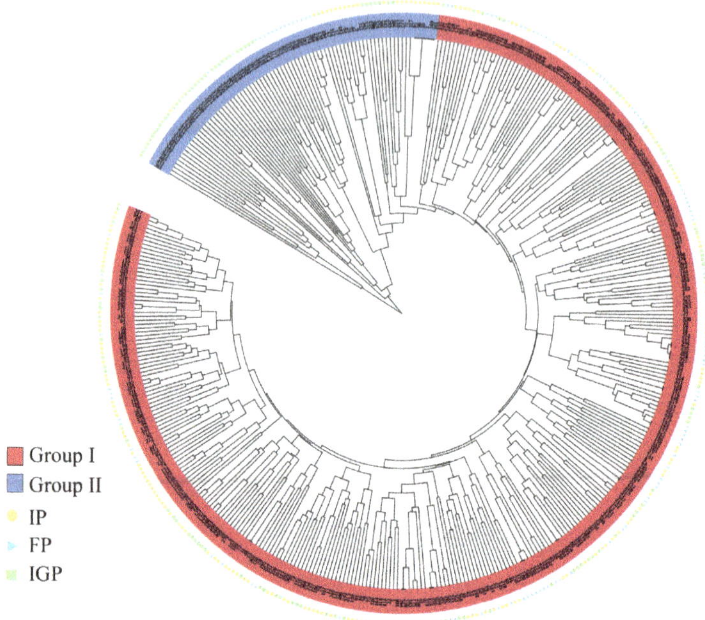

Figure 5. UPGMA cluster analysis of the *P. elliottii* provenances based on the genetic distance.

4. Discussion

4.1. Development and Cross-Species Applicability of P. elliottii Primers

The limited availability of SSR primers has hindered molecular-level research on *P. elliottii* populations. In this study, we developed 19 polymorphic SSR loci based on the *P. elliottii* transcriptome and conducted cross-species applicability analyses across three pine species. The results demonstrated that the SSR markers developed for *P. elliottii* exhibited high cross-species applicability in other *Pinus* species, such as *P. taeda* and *P. massoniana*, with a 100% amplification rate and a high proportion of polymorphic loci—89.47% in *P. taeda* and 100% in *P. massoniana*. This high level of cross-species applicability is likely due to the conserved nature of the EST sequences compared to the non-coding DNA, along with the low molecular evolution rates within the *Pinus* genus, which facilitate the cross-transferability of SSRs and other molecular markers [37,38]. For instance, 14 EST-SSR markers developed for Aleppo pine (*Pinus halepensis*) showed high applicability in Mediterranean pines (*Pinus halepensis*) [39], and 184 EST-SSR markers developed from the lodgepole pine transcriptome were effectively amplified across different Pinus species, with amplification rates ranging from 54% to 73%, including a 65% amplification rate in *P. elliottii* [40]. Similarly, 20 EST-SSR markers developed for *Larix principis-rupprechtii* produced clear and stable bands across three related *Larix* species [41]. In this study, the SSR primers developed for *P. elliottii* demonstrated 100% cross-species applicability and a high proportion of polymorphic loci in *P. massoniana*, consistent with the findings from previous studies. These EST-SSR markers can also be employed in the future for interspecific hybrid identification, comparative mapping, and phylogenetic relationship evaluation among *P. elliottii* and its related species.

The PIC is a crucial genetic parameter often used to assess the discriminatory power of primers and the reliability of the information they provide in genetic diversity studies [42]. According to the definition of polymorphism by Botstein et al., the SSR primers selected in this study had an average PIC value of 0.411, with six loci exhibiting PIC values greater than 0.5, indicating a high level of polymorphism [43]. This suggests that the 19 SSR primers exhibit good polymorphism, effectively reflecting the genetic diversity of the materials. Therefore, these primers can provide technical support for the identification of *P. elliottii* germplasm, variety breeding, and the construction of genetic fingerprints. In comparison to previously developed SSR markers from our research group (average PIC = 0.349), the polymorphism of the primers used in this study was improved and the number of analyzed populations was expanded.

4.2. Genetic Diversity of P. elliottii Populations

Genetic diversity is a crucial factor in assessing a species' ability to adapt to environmental changes. Enhancing the genetic base of a population can increase its adaptability and resilience, thereby improving its overall survival potential [17]. The level of genetic diversity within populations can be reflected by the *He* and the *I* [44].

In this study, 19 pairs of SSR primers were used to amplify three *P. elliottii* provenances, resulting in $I = 0.862$ and $He = 0.457$. These values were higher than those reported for Scots pine ($He = 0.167$; $I = 0.253$) [45], *P. yunnanensis* var. *tenuifolia* ($He = 0.342$; $I = 0.488$) [46], and *P. taiwanensis* Hayata var. *damingshanensis* ($He = 0.40$; $I = 0.62$) [47], but lower than those for *P. massoniana* ($He = 0.5438$; $I = 0.8696$) [11], Chinese pine ($He = 0.604$; $I = 1.230$) [48], and Korean pine ($He = 0.521$; $I = 1.019$) [49]. These results indicate that *P. elliottii* exhibits a moderate to high level of genetic diversity compared to other conifer species.

Some studies suggest that while breeding practices can effectively increase the genetic gains of improved populations, they may also reduce the genetic diversity of these populations due to intensified human selection pressure [50]. However, empirical research has shown that the impact of artificial selection on genetic diversity is not significant. For instance, comparative analyses of the genetic diversity between improved populations obtained through phenotypic selection and their natural source populations in white spruce (*Picea glauca*) found only a slight, non-significant reduction in diversity in the improved populations [51]. Similarly, genetic diversity analyses of first- and second-generation improved seed orchards of *Pinus taeda* and first- to third-generation improved seed orchards of *Larix principis-rupprechtii* revealed that the genetic diversity remained relatively stable across different breeding generations, with only minor population differentiation [52,53]. Even in the fourth-generation breeding population of *Cunninghamia lanceolata*, high genetic diversity was maintained [54], and studies on *Picea sitchensis* [55], *Pinus taeda* [56], and *Larix kaempferi* [50] have found that the genetic diversity of improved populations can sometimes exceed that of their source populations.

In this study, no significant differences were found in the *I*, *Ho*, and *He* among the three provenances of slash pine. However, the genetic diversity of the FP and IGP was slightly higher than that of the IP. These findings align with the results observed in *Larix kaempferi* [50], *Picea sitchensis* [55], and *Pinus taeda* [56], suggesting that the breeding strategies and measures applied during the genetic improvement of slash pine in Jiangxi have been relatively effective. As an introduced species without natural germplasm resources in China, the genetic diversity of slash pine relies on the diversity of the introduced germplasm. To further enhance the genetic diversity of breeding populations, it may be advantageous to introduce superior individuals from native populations with distinct genetic backgrounds or to select elite offspring from already introduced families. This approach would help to further broaden the genetic base of the breeding populations.

4.3. Genetic Differentiation and Gene Flow in P. elliottii Populations

Understanding the genetic differentiation and structure of populations is fundamental for developing species conservation strategies and constructing plant breeding populations. Research suggests that a *Fst* value of less than 0.05 indicates low genetic differentiation, $0.05 < Fst < 0.15$ indicates moderate differentiation, $0.15 < Fst < 0.25$ indicates relatively high differentiation, and $Fst > 0.25$ indicates high genetic differentiation [57]. In this study, the *Fst* value for *P. elliottii* was 0.016, which is less than 0.05, indicating low genetic differentiation among the populations. Additionally, the AMOVA analysis showed that the within-provenance variation (97.9%) was much greater than the between-provenance variation (2.10%). This, combined with the genetic differentiation analysis, demonstrates that *P. elliottii* has significantly higher genetic diversity within populations than between them, suggesting that selection efforts should focus more on individual trees within populations. This phenomenon has also been observed in other conifer species, such as Korean pine [14], *P. massoniana* [11], Norway spruce [58], *Pinus yunnanensis* var. *tenuifolia* [46], and *Cunninghamia lanceolata* [59]. The likely cause of this pattern is the increased frequency of gene exchange between different population germplasm resources [60].

The gene flow (*Nm*) is a critical factor influencing genetic differentiation among populations. Hamrick et al. suggested that when $Nm > 4$, the gene exchange between populations was relatively sufficient to counteract genetic drift and prevent genetic differentiation among populations [61]. In this study, the overall gene flow among provenances was 15.715, which aligns with the observed low level of genetic differentiation ($Fst = 0.016$). This high level of *Nm* has inhibited population differentiation and inbreeding, promoting individual variation and leading to a relatively uniform genetic background across populations. These findings are consistent with the results of the genetic structure analysis and variance analysis. Two primary factors may contribute to this pattern. First, it is related to the biological characteristics of *P. elliottii*. The species produces winged seeds and is wind-pollinated, with pollen-containing air sacs that allow it to travel relatively long distances [62,63]. Second, there has been frequent human-mediated exchange of germplasm materials between different populations over a long period, with most population materials being confined to Jiangxi Province [54]. This has resulted in low genetic diversity and a lack of significant genetic differentiation among populations. In the future selection of *P. elliottii* breeding materials, the focus should be on selecting and preserving superior individuals within each population. Additionally, selecting elite trees from major *P. elliottii* introduction areas could help expand the range of resource collection, enrich the diversity of genetic resources, and improve the breeding efficiency.

4.4. Genetic Structure and Clustering Analysis of P. elliottii Populations

Population structure analysis in plants primarily utilizes three different methods: STRUCTURE analysis, UPGMA clustering, and PCoA analysis. PCoA and UPGMA offer a visual representation of the distribution relationships among materials, while STRUCTURE analysis provides insights into the genetic background and gene flow within populations [64,65]. These three methods complement and corroborate each other, facilitating a comprehensive understanding of the genetic differences among materials. In this study, STRUCTURE analysis revealed that the 417 *P. elliottii* samples could be divided into two subgroups. Individuals from different sources clustered together within the same genetic structure, indicating that most samples had relatively homogeneous genetic backgrounds. Using a Q value threshold of 0.6 for classification, only seven samples had Q values below 0.6, suggesting that these samples have a more complex genetic structure and evidence of gene flow. Therefore, caution should be exercised when selecting these samples as breeding parents. Analysis using the UPGMA clustering also grouped the samples into two main clusters, and with few exceptions, the clustering patterns were consistent with the results of both the PCoA and STRUCTURE analyses.

5. Conclusions

This study examined the genetic diversity and population structure of 417 *P. elliottii* samples from three provenances using 19 polymorphic SSR markers. The results showed that *P. elliottii* exhibited moderate to high genetic diversity compared to other conifer species. The gene flow and genetic differentiation analyses indicated that the gene exchange among the provenances was relatively extensive, limiting genetic differentiation and resulting in low overall genetic differentiation. The molecular variance analysis revealed that most genetic variation occurred within populations. Therefore, breeding efforts should focus on selecting superior individuals from genetically diverse populations. The population structure analysis revealed that individuals from different sources were not strictly clustered according to their origins but were distributed across two subgroups, with seven samples classified into a mixed group, indicating complex genetic relationships. The presence of redundant genetic diversity among the 417 germplasm samples suggests that future work should aim to remove genetic redundancies. Overall, these findings provide valuable guidance for the efficient utilization of high-resin-yielding *P. elliottii* germplasm, the development of core germplasm, and the selection of hybrid parents for breeding new germplasm.

Author Contributions: Conceptualization, M.Y. and M.L.; methodology, M.Y.; software, R.H.; validation, M.Y., R.H. and W.H.; formal analysis, H.X. and W.H.; investigation, H.X., W.X. and X.L.; resources, M.Y. and M.L.; data curation, R.H., W.X. and T.C.; writing—original draft preparation, M.Y.; writing—review and editing, M.L.; supervision, M.L.; project administration, M.L. All authors have read and agreed to the published version of the manuscript.

Funding: This research was funded by several programs, including the Jiangxi Province Science Foundation for Youths (No. 20212BAB215013), the Science and Technology Research Project of Education Department of Jiangxi Province (No. GJJ200407), the National Natural Science Foundation of China (No. 32160385 and 32460385), and the Jiangxi Forestry Science and Technology Innovation Project (No. 202311).

Institutional Review Board Statement: Not applicable.

Informed Consent Statement: Not applicable.

Data Availability Statement: Data are contained within the article.

Acknowledgments: We thank the editor and reviewers for their constructive comments and advice on our manuscript.

Conflicts of Interest: The authors declare no conflicts of interest.

References

1. Susaeta, A.; Peter, G.F.; Hodges, A.W.; Carter, D.R. Oleoresin tapping of planted slash pine (*Pinus elliottii* Engelm. var. *elliottii*) adds value and management flexibility to landowners in the southern United States. *Biomass Bioenergy* **2014**, *68*, 55–61.
2. Neis, F.A.; de Costa, F.; de Almeida, M.R.; Colling, L.C.; de Oliveira Junkes, C.F.; Fett, J.P.; Fett-Neto, A.G. Resin exudation profile, chemical composition, and secretory canal characterization in contrasting yield phenotypes of *Pinus elliottii* Engelm. *Ind. Crop. Prod.* **2019**, *132*, 76–83. [CrossRef]
3. Lai, M.; Dong, L.M.; Su, R.F.; Zhang, L.; Jia, T.; Chen, T.X.; Yi, M. Needle functional features in contrasting yield phenotypes of slash pine at three locations in southern China. *Ind. Crop. Prod.* **2023**, *206*, 117613. [CrossRef]
4. Ding, X.; Li, Y.; Zhang, Y.; Diao, S.; Luan, Q.; Jiang, J. Genetic analysis and elite tree selection of the main resin components of slash pine. *Front. Plant Sci.* **2023**, *14*, 1079952. [CrossRef] [PubMed]
5. Lai, M.; Zhang, L.; Lei, L.; Liu, S.; Jia, T.; Yi, M. Inheritance of resin yield and main resin components in *Pinus elliottii* Engelm. at three locations in southern China. *Ind. Crop. Prod.* **2020**, *144*, 112065. [CrossRef]
6. Yi, M.; Zhang, L.; Cheng, Z.; Hu, R.; Gao, Y.; Jin, C.; Yuan, S.; Sun, S.; Lai, M. Identification of key genes for oleoresin biosynthesis in high and low oleoresin-yielding slash pine based on transcriptome analysis. *Forests* **2022**, *13*, 1337. [CrossRef]
7. Dungey, H.S. Pine hybrids—A review of their use performance and genetics. *For. Ecol. Manage.* **2001**, *148*, 243–258. [CrossRef]
8. Sun, M.S.; Feng, Y.H.; Jia, J.; Yang, Z.Q. Fertility of different interspecific hybrid types of pines. *Guihaia* **2021**, *41*, 1270–1279.
9. Luan, Q.; Lu, P.; Xiao, F.; Jiang, J.; Yu, M. Investigation on the damage of *Pinus elliottii* in the freezing rain and snow area and the analysis on the reason. *Sci. Silv. Sin.* **2008**, *44*, 50–55.
10. Zhou, B.; Gu, L.; Ding, Y.; Shao, L.; Wu, Z.; Yang, X.; Li, C.; Li, Z.; Wang, X.; Cao, Y.; et al. The great 2008 Chinese ice storm: Its socioeconomic-ecological impact and sustainability lessons learned. *Bull. Am. Meteorol. Soc.* **2011**, *92*, 47–60. [CrossRef]
11. Mei, L.; Wen, X.; Fan, F.; Yang, Z.; Xie, W.; Hong, Y. Genetic diversity and population structure of masson pine (*Pinus massoniana* Lamb.) superior clones in South China as revealed by EST-SSR markers. *Genet. Resour. Crop. Evol.* **2021**, *68*, 1987–2002. [CrossRef]
12. Sheller, M.; Tóth, E.G.; Ciocîrlan, E.; Mikhaylov, P.; Kulakov, S.; Kulakova, N.; Melnichenko, N.; Ibe, A.; Sukhikh, T.; Curtu, A.L. Genetic diversity and population structure of scots pine (*Pinus sylvestris* L.) in Middle Siberia. *Forests* **2023**, *14*, 119. [CrossRef]
13. Lu, M.M.; Krutovsky, K.V.; Nelson, C.D.; Koralewski, T.E.; Byram, T.D.; Loopstra, C.A. Exome genotyping, linkage disequilibrium and population structure in loblolly pine (*Pinus taeda* L.). *BMC Genom.* **2016**, *17*, 730. [CrossRef]
14. Yan, P.; Xie, Z.; Feng, K.; Qiu, X.; Zhang, L.; Zhang, H. Genetic diversity analysis and fingerprint construction of Korean pine (*Pinus koraiensis*) clonal seed orchard. *Front. Plant Sci.* **2023**, *13*, 1079571. [CrossRef]

15. Gramazio, P.; Plesa, I.M.; Truta, A.M.; Sestras, A.F.; Vilanova, S.; Plazas, M.; Vicente, O.; Boscaiu, M.; Prohens, J.; Sestras, R.E. Highly informative SSR genotyping reveals large genetic diversity and limited differentiation in European larch (*Larix decidua*) populations from Romania. *Turk. J. Agric. For.* **2018**, *42*, 165–175. [CrossRef]
16. Rana, J.C.; Chahota, R.K.; Sharma, V.; Rana, M.; Verma, N.; Verma, B.; Sharma, T.R. Genetic diversity and structure of Pyrus accessions of Indian Himalayan region based on morphological and SSR markers. *Tree Genet. Genomes* **2015**, *11*, 821. [CrossRef]
17. Porth, I.; El-Kassaby, Y.A. Assessment of the genetic diversity in forest tree populations using molecular markers. *Diversity* **2014**, *6*, 283–295. [CrossRef]
18. Grover, A.; Sharma, P.C. Development and use of molecular markers: Past and present. *Crit. Rev. Biotechnol.* **2016**, *36*, 290–302. [CrossRef]
19. Powell, W.; Machray, G.C.; Provan, J. Polymorphism revealed by simple sequence repeats. *Trends Plant Sci.* **1996**, *1*, 215–222. [CrossRef]
20. Carletti, G.; Cattivelli, L.; Vietto, L.; Nervo, G. Multiallelic and multilocus simple sequence repeats (SSRs) to assess the genetic diversity of a Salix spp. germplasm collection. *J. For. Res.* **2021**, *32*, 263–271. [CrossRef]
21. Kavaliauskas, D.; Danusevičius, D.; Baliuckas, V. New insight into genetic structure and diversity of Scots pine (*Pinus sylvestris* L.) populations in Lithuania based on nuclear, chloroplast and mitochondrial DNA markers. *Forests* **2022**, *13*, 1179. [CrossRef]
22. Zhou, B.; Zhang, Z.; Li, Y.; Ma, Y.; Zhang, S.; Niu, S.; Li, Y. Genetic diversity, genetic structure, and germplasm source of Chinese pine in North China. *Eur. J. For. Res.* **2023**, *142*, 183–195. [CrossRef]
23. de Oliveira Junkes, C.F.; de Araújo Júnior, A.T.; de Lima, J.C.; de Costa, F.; Füller, T.; de Almeida, M.R.; Neis, F.A.; da Silva Rodrigues-Correa, K.C.; Fett, J.P.; Fett-Neto, A.G. Resin tapping transcriptome in adult slash pine (*Pinus elliottii* var. *elliottii*). *Ind. Crop. Prod.* **2019**, *139*, 111545. [CrossRef]
24. Yi, M.; Zhang, L.; Lei, L.; Cheng, Z.; Sun, S.; Lai, M. Analysis of SSR information in transcriptome and development of EST-SSR molecular markers in *Pinus elliottii* Engelm. *J. Nanjing For. Univ.* **2020**, *44*, 75.
25. Nelson, C.D.; Nance, W.L.; Doudrick, R.L. A partial genetic linkage map of slash pine (*Pinus elliottii* Engelm. var. *elliottii*) based on random amplified polymorphic DNAs. *Theor. Appl. Genet.* **1993**, *87*, 145–151.
26. Acosta, J.J.; Fahrenkrog, A.M.; Neves, L.G.; Resende, M.F.; Dervinis, C.; Davis, J.M.; Holliday, J.A.; Kirst, M. Exome resequencing reveals evolutionary history, genomic diversity, and targets of selection in the conifers *Pinus taeda* and *Pinus elliottii*. *Genome Biol. Evol.* **2019**, *11*, 508–520. [CrossRef]
27. Westbrook, J.W.; Chhatre, V.E.; Wu, L.S.; Chamala, S.; Neves, L.G.; Muñoz, P.; Martínez-García, P.J.; Neale, D.B.; Kirst, M.; Mockaitis, K.; et al. A consensus genetic map for *Pinus taeda* and *Pinus elliottii* and extent of linkage disequilibrium in two genotype-phenotype discovery populations of *Pinus taeda*. *G3 Genes Genomes Genet.* **2015**, *5*, 1685–1694. [CrossRef]
28. Untergasser, A.; Cutcutache, I.; Koressaar, T.; Ye, J.; Faircloth, B.; Remm, M.; Rozen, S. Primer3—New capabilities and interfaces. *Nucleic Acids Res.* **2012**, *40*, 115. [CrossRef]
29. Holland, M.; Parson, W. GeneMarker® HID: A reliable software tool for the analysis of forensic STR data. *J. Forensic Sci.* **2011**, *56*, 29–35. [CrossRef]
30. Fan, W.; Gai, H.; Sun, X.; Yang, A.; Zhang, Z.; Ren, M. DataFormater, a software for SSR data formatting to develop population genetics analysis. *Mol. Plant Breed.* **2016**, *14*, 265–270.
31. Peakall, R.O.D.; Smouse, P.E. GENALEX 6: Genetic analysis in Excel. Population genetic software for teaching and research. *Mol. Ecol. Notes* **2006**, *6*, 288–295. [CrossRef]
32. Yeh, F.C.; Yang, R.C.; Boyle, T.B.J.; Ye, Z.H.; Mao, J.X. Popgene, the user friendly shareware for population genetic analysis. In *Molecular Biology and Biotechnology Centre*; University of Alberta: Edmonton, AB, Canada, 1997.
33. Liu, K.; Muse, S.V. PowerMarker: An integrated analysis environment for genetic marker analysis. *Bioinformatics* **2005**, *21*, 2128–2129. [CrossRef]
34. Excoffier, L.; Lischer, H.E.L. Arlequin suite ver 3.5: A new series of programs to perform population genetics analyses under Linux and Windows. *Mol. Ecol. Resour.* **2010**, *10*, 564–567. [CrossRef]
35. Evanno, G.; Regnaut, S.; Goudet, J. Detecting the number of clusters of individuals using the software STRUCTURE: A simulation study. *Mol. Ecol.* **2005**, *14*, 2611–2620. [CrossRef]
36. Tamura, K.; Dudley, J.; Nei, M.; Kumar, S. MEGA4: Molecular evolutionary genetics analysis (MEGA) software version 4.0. *Mol. Biol. Evol.* **2007**, *24*, 1596–1599. [CrossRef]
37. Neale, D.B. Genomics to tree breeding and forest health. *Curr. Opin. Genet. Dev.* **2007**, *17*, 539–544. [CrossRef]
38. Neale, D.B.; Ingvarsson, P.K. Population, quantitative and comparative genomics of adaptation in forest trees. *Curr. Opin. Plant Biol.* **2008**, *11*, 149–155. [CrossRef]
39. Leonarduzzi, C.; Spanu, I.; Labriola, M.; González-Martínez, S.C.; Piotti, A.; Vendramin, G.G. Development and characterization of three highly informative EST-SSR multiplexes for *Pinus halepensis* Mill. and their transferability to other Mediterranean pines. *Plant Mol. Biol. Rep.* **2016**, *34*, 993–1002. [CrossRef]
40. Lesser, M.R.; Parchman, T.L.; Buerkle, C.A. Cross-species transferability of SSR loci developed from transciptome sequencing in lodgepole pine. *Mol. Ecol. Resour.* **2012**, *12*, 448–455. [CrossRef]
41. Dong, M.; Wang, Z.; He, Q.; Zhao, J.; Fan, Z.; Zhang, J. Development of EST-SSR markers in *Larix principis-rupprechtii* Mayr and evaluation of their polymorphism and cross-species amplification. *Trees* **2018**, *32*, 1559–1571. [CrossRef]

42. Uzun, A.; Yesiloglu, T.; Polat, I.; Aka-Kacar, Y.; Gulsen, O.; Yildirim, B.; Tuzcu, O.; Tepe, S.; Canan, I.; Anil, S. Evaluation of genetic diversity in lemons and some of their relatives based on SRAP and SSR markers. *Plant Mol. Biol. Rep.* **2011**, *29*, 693–701. [CrossRef]
43. Botstein, D.; White, R.L.; Skolnick, M.; Davis, R.W. Construction of a genetic linkage map in man using restriction fragment length polymorphisms. *Am. J. Hum. Genet.* **1980**, *32*, 314. [PubMed]
44. Schmidt, T.L.; Jasper, M.E.; Weeks, A.R.; Hoffmann, A.A. Unbiased population heterozygosity estimates from genome-wide sequence data. *Methods Ecol. Evol.* **2021**, *12*, 1888–1898. [CrossRef]
45. Vasilyeva, Y.; Chertov, N.; Nechaeva, Y.; Sboeva, Y.; Pystogova, N.; Boronnikova, S.; Kalendar, R. Genetic structure, differentiation and originality of *Pinus sylvestris* L. populations in the east of the East European Plain. *Forests* **2021**, *12*, 999. [CrossRef]
46. Yang, Z.; Feng, Y.; Wu, D. Analysis of genetic diversity of *Pinus yunnanensis* var. *tenuifolia* nature populations by SSR marker. *Guihaia* **2014**, *34*, 10–14.
47. Luo, Q.; Feng, Y.; Wu, D.; Yang, Z. Genetic diversity of *Pinus taiwanensis* var. *damingshanensis* natural populations by SSR markers. *Guihaia* **2022**, *42*, 1367–1373.
48. Yang, B.; Niu, S.; El-Kassaby, Y.A.; Li, W. Monitoring genetic diversity across *Pinus tabuliformis* seed orchard generations using SSR markers. *Can. J. For. Res.* **2021**, *51*, 1534–1540. [CrossRef]
49. Li, X.; Zhao, M.; Xu, Y.; Li, Y.; Tigabu, M.; Zhao, X. Genetic diversity and population differentiation of *Pinus koraiensis* in China. *Horticulturae* **2021**, *7*, 104. [CrossRef]
50. Du, C.; Sun, X.; Xie, Y.; Hou, Y. Genetic diversity of *Larix kaempferi* populations with different levels of improvement in northern subtropical region. *Sci. Silvae Sin.* **2021**, *57*, 68–76.
51. Fageria, M.S.; Rajora, O.P. Effects of silvicultural practices on genetic diversity and population structure of white spruce in Saskatchewan. *Tree Genet. Genomes* **2014**, *10*, 287–296. [CrossRef]
52. Chhatre, V.E.; Byram, T.D.; Neale, D.B.; Wegrzyn, J.L.; Krutovsky, K.V. Genetic structure and association mapping of adaptive and selective traits in the east Texas loblolly pine (*Pinus taeda* L.) breeding populations. *Tree Genet. Genomes* **2013**, *9*, 1161–1178. [CrossRef]
53. Yu, D.; Yuan, D.; Zhang, D.; Fan, Y.; Li, D.; Zhang, H.; Zhang, J. Genetic diversity of *Larix principisrupprechtii* Mayr. seed orchard among generations. *J. Plant Genet. Resour.* **2014**, *15*, 940–947.
54. Jing, Y.; Bian, L.; Zhang, X.; Zhao, B.; Zheng, R.; Su, S.; Ye, D.; Zheng, X.; El-Kassaby, Y.A.; Shi, J. Genetic diversity and structure of the 4th cycle breeding population of Chinese fir (*Cunninghamia lanceolata* (lamb.) hook). *Front. Plant Sci.* **2023**, *14*, 1106615. [CrossRef] [PubMed]
55. Chaisurisri, K.; El-Kassaby, Y.A. Genetic diversity in a seed production population vs. natural populations of Sitka spruce. *Biodivers. Conserv.* **1994**, *3*, 512–523. [CrossRef]
56. Isik, F.; McKeand, S.E. Fourth cycle breeding and testing strategy for *Pinus taeda* in the NC State University Cooperative Tree Improvement Program. *Tree Genet. Genomes* **2019**, *15*, 70. [CrossRef]
57. Pearse, D.E.; Crandall, K.A. Beyond F ST: Analysis of population genetic data for conservation. *Conserv. Genet.* **2004**, *5*, 585–602. [CrossRef]
58. Stojnić, S.V.; Avramidou, E.; Fussi, B.; Westergren, M.; Orlović, S.; Matović, B.; Trudić, B.; Kraigher, H.A.; Aravanopoulos, F.; Konnert, M. Assessment of genetic diversity and population genetic structure of Norway spruce (*Picea abies* L.) Karsten at Its southern lineage in Europe. Implications for conservation of forest genetic resources. *Forests* **2019**, *10*, 258. [CrossRef]
59. Lin, E.; Zhuang, H.; Yu, J.; Liu, X.; Huang, H.; Zhu, M.; Tong, Z. Genome survey of Chinese fir (*Cunninghamia lanceolata*): Identification of genomic SSRs and demonstration of their utility in genetic diversity analysis. *Sci. Rep.* **2020**, *10*, 4698. [CrossRef]
60. Ithnin, M.; The, C.K.; Ratnam, W. Genetic diversity of *Elaeis oleifera* (HBK) Cortes populations using cross species SSRs: Implication's for germplasm utilization and conservation. *BMC Genet.* **2017**, *18*, 37. [CrossRef]
61. Hamrick, J.L.; Godt, M.J.W. Allozyme diversity in plants. In *Population Genetics, Breeding and Genetic Resources*, 2nd ed.; Brown, A.D.H., Clegg, M.T., Kahler, A.L., Weir, B.S., Eds.; Sinauer & Associates: Sunderland, MA, USA, 1989; pp. 43–63.
62. Petit, R.J.; Duminil, J.; Fineschi, S.; Hampe, A.; Salvini, D.; Vendramin, G.G. Invited review: Comparative organization of chloroplast, mitochondrial and nuclear diversity in plant populations. *Mol. Ecol.* **2005**, *14*, 689–701. [CrossRef]
63. Tong, Y.; Lewis, B.J.; Zhou, W.; Mao, C.; Wang, Y.; Zhou, L.; Yu, D.; Dai, L.; Qi, L. Genetic diversity and population structure of natural *Pinus koraiensis* populations. *Forests* **2019**, *11*, 39. [CrossRef]
64. Wang, S. Genetic diversity and population structure of the endangered species *Paeonia decomposita* endemic to China and implications for its conservation. *BMC Plant Biol.* **2020**, *20*, 510. [CrossRef] [PubMed]
65. Deng, L.; Liu, Q.; Zhou, Z.; Gao, K.; Luo, D. Genetic diversity analysis and core collection of pinewood nematodiasis-resistant *Pinus massoniana* germplasm resources. *J. Zhejiang Agric. For. Univ.* **2024**, *41*, 67–78.

life

MDPI

Article

Effects of Land Use Type Transformation on the Structure and Diversity of Soil Bacterial Communities

Henian Hua [1,†], Xin Sui [2,†], Yanan Liu [1], Xu Liu [1], Qiuyang Chang [1], Ruiting Xu [1], Mengsha Li [3,*] and Liqiang Mu [1,*]

[1] Key Laboratory of Sustainable Forest Ecosystem Management-Ministry of Education, School of Forestry, Northeast Forestry University, Harbin 150040, China; hhn54553627@nefu.edu.cn (H.H.); lynm@nefu.edu.cn (Y.L.); liuxu19981209@nefu.edu.cn (X.L.); 1413085125@nefu.edu.cn (Q.C.); xrt@nefu.edu.cn (R.X.)

[2] Engineering Research Center of Agricultural Microbiology Technology, Ministry of Education & Heilongjiang Provincial Key Laboratory of Ecological Restoration and Resource Utilization for Cold Region & Key Laboratory of Microbiology, College of Heilongjiang Province & School of Life Sciences, Heilongjiang University, Harbin 150080, China; xinsui_cool@126.com

[3] Institute of Nature and Ecology, Heilongjiang Academy of Sciences, Harbin 150040, China

* Correspondence: lms19861004@163.com (M.L.); mlq0417@163.com (L.M.)

† These authors contributed equally to this work.

Abstract: Soil microbiota are significantly influenced by their microenvironments. Therefore, to understand the impacts of various land use patterns on the diversity and composition of soil bacterial communities, this study focused on three typical land use types—NF (natural forest), AF (artificial forests), and FL (farmland)—in the Heilongjiang Central Station Black-billed Capercaillie National Nature Reserve, located in the southwestern part of Heihe City, Heilongjiang Province, China. Using high-throughput sequencing of the 16S rRNA gene, we examined the soil bacterial community structures in these different land use types and explored their correlation with soil environmental factors. The following were our main observations: (1) Significant variations in soil chemical properties among different land use patterns were observed. In artificial forests, total nitrogen (TN), alkali hydrolyzed nitrogen (AN), total phosphorus (TP), and available phosphorus (AP) were higher compared to farmland and significantly higher than those in natural forests. Furthermore, the organic carbon content (SOC) in natural forests was higher than in artificial forests and significantly higher than in farmland. (2) Comparative analysis using the Shannon and Simpson indices revealed that bacterial community diversity was higher in artificial forests than in natural forests, which was significantly higher than in farmland. (3) The effect of different land use types on soil bacterial community structure was not significant. The three land types were dominated by Proteobacteria, Acidobacteria, and Actinobacteria. Proteobacteria exhibited a higher relative abundance in farmland and artificial forests compared to natural forests, whereas Actinobacteria exhibited the lowest relative abundance in natural forests. (4) Redundancy analysis (RDA) revealed that SOC, TN, AN, and AP were key environmental factors influencing the microbial communities of soil. Collectively, our findings demonstrated that land use practices can significantly alter soil nutrient levels, thereby influencing the structure of bacterial communities.

Keywords: land use type; soil microorganisms; community structure; diversity

1. Introduction

Changes in land use have a profound impact on terrestrial ecosystems and biogeochemical cycles, thereby altering soil properties and fertility [1]. These alterations, in turn, influence the diversity and composition of soil microbial communities [2]. Soil microorganisms are a crucial component of soil ecosystems and are intricately linked to plant growth and soil biogeochemical cycling [3]. Moreover, the composition and structure of soil microbial communities significantly shape soil structure and functionality [4,5]. Therefore, the impact of different land use modes on the structure and functionality of soil ecosystems can be predicted by observing structural changes in soil microorganisms [6].

Different land use practices induce alterations in vegetation coverage and soil physicochemical properties. These changes are closely linked to the environmental heterogeneity of soil and represent pivotal factors influencing the structure of the microbial communities of soil [7,8]. For example, ref. [9] observed significant variations in soil bacterial diversity and community distribution across different land use types in the Chongqing Yangtze River Basin. Furthermore, the authors noted a consistent relationship between changes in bacterial diversity and alterations in soil physicochemical properties. Additionally, α-diversity is commonly utilized as a basis to analyze soil microbial diversity. Li Jinbiao et al. [10] found that alterations in land use types lead to changes in soil chemical properties, significantly impacting the composition and diversity of soil bacterial communities. The chemical properties most affected by changes in land use types were soil organic carbon content, followed by total phosphorus, total nitrogen, and total potassium. Moreover, Wu Yining et al. analyzed microbial community composition and functionality in marsh wetlands, meadow wetlands, forests, and farmlands and reported that soil bacterial abundance in forested areas significantly surpassed that of other land use patterns. Additionally, the authors noted that total nitrogen significantly influenced both the diversity and abundance of soil microbial communities [11]. Multiple studies have characterized the effects of anthropogenic activities on the diversity and composition of soil microbial communities, including agriculture, urban development, industry, and pesticide use [12,13]. Nevertheless, the complex environmental factors and management practices involved in different land use practices may have synergistic effects on the microbial composition of soil, and therefore more comprehensive studies are needed to understand this complex relationship. Understanding the effects of land use practices on soil microorganisms can thus provide a theoretical and practical basis for the management and conservation of soil ecosystems.

Heilongjiang Province boasts a rich agricultural history and fertile land, making it one of China's foremost centers for commodity grain production. The soil's physicochemical properties and fertility in this region are of paramount importance for the advancement of agriculture [14]. The Heilongjiang Central Station Black-billed Capercaillie National Nature Reserve is located in the transitional zone between the Greater and Lesser Khingan Mountains. The primary dominant community within the reserve comprises the Dahurian larch forests. Moreover, the Central Station stands as one of the oldest regions in Northeast China in terms of historical development and utilization. However, it is also considered an ecologically fragile environment. Due to population growth, this region has resorted to deforestation and land clearance to enhance agricultural yields and address food security concerns. Human intervention has resulted in the degradation of primary forests, leading to the emergence of extensive secondary vegetation, such as birch and Mongolian oak forests. These alterations in forest types not only impact ecosystem functionalities but also exert a notable influence on the soil ecosystem [15]. Furthermore, the conversion of natural forests into agricultural fields has led to the loss of natural vegetative cover, leaving the soil surface exposed and causing severe detriment to the soil ecosystem.

Based on the aforementioned background, our study sought to comparatively analyze the bacterial community composition and diversity across three distinct land use patterns within the Central Station area (natural forest, plantation forest, farmland) using a high-throughput sequencing approach. Particularly, this study aimed to establish a scientific basis for the preservation of soil fertility, nurturing practices, and the conservation of soil microbial diversity to enable ecological reconstruction and sustainable land resource utilization in the study area.

2. Materials and Methods

2.1. Study Site

The study site was situated in the Heilongjiang Central Station Black-billed Capercaillie National Nature Reserve, located in the southwestern part of Heihe City, Heilongjiang Province. This site lies within the transitional zone between the southwestern foothills of the Lesser Khingan Mountains and the Songnen Plain, encompassing a total area of 988.6 km^2. The region is located within the temperate continental monsoon climate zone, characterized by prolonged and harsh winters as well as brief and cool summers. The average annual temperature is −0.5 °C, with a frost-free period lasting 121 days. The annual average rainfall is 476.33 mm, and the average relative humidity is 69.2%. The nature reserve features a typical alpine forest ecosystem, encompassing forests, shrubs, wetlands, and meadows. The forests include coniferous forests, coniferous and broad-leaved mixed forests, and broad-leaved forests, boasting a high forest coverage of 82.4% [14]. The main plant species in the study area include *Larix gmelinii* larch, *Betula platyphyllabirch*, black birch, *Betula dahurica*, *Quercus mongolica*, Mongolian oak, Amur linden (*Tilia amurensis*), diamond willow (*Chosenia arbutifolia*), aspen (*Populus davidiana*), bird cherry (*Prunus padus*), *Salix raddeana*, and alder trees (*Alnus mandshurica*).

2.2. Sample Collection

In June 2022, three typical land use types: plantation *Larix gmelinii* forest, natural forest (mixed forests of *Larix gmelinii* forest, *Betula platyphylla* forest, and *Betula dahurica* forest), and agricultural land (soybean cultivation) were selected in this nature reserve. Each land use type included three independent replicates, and each replicate also included three standard quadrats, so a total of nine standard quadrats were established. Each quadrat set is 50 × 50 m. Soil samples were collected at a depth of 0–20 cm using a soil shovel with a five-point sampling method, which were mixed as one soil sample. After collection, the soil samples were first cleaned of plant debris and then passed through a 2 mm sieve. The processed soil samples were then sealed in ziplock bags and stored at −20 °C in the laboratory for subsequent analysis.

2.3. Analysis of Soil Physicochemical Properties

Soil total nitrogen was determined using the Kjeldahl method, whereas available nitrogen was assessed using the alkaline hydrolysis diffusion method. Soil total phosphorus was determined via the sulfuric acid-perchloric acid digestion method followed by the molybdenum-antimony colorimetric technique, while available phosphorus content was quantified using the bicarbonate method. The total potassium content of the soil was analyzed using the sodium hydroxide fusion-flame photometer method. Available potassium was quantified using ammonium acetate extraction, followed by flame photometry. Soil organic carbon content was measured with additional heating using an oil bath.

2.4. Sample DNA Extraction

Genomic DNA was extracted using the E.Z.N.A.® Soil Kit (Omega Bio-tek, Norcross, GA, USA). The quantity and integrity of the extracted DNA were then preliminarily examined using 1% agarose gel electrophoresis. Afterward, the DNA was precisely quantified using the QuantiFluor™—ST Blue Fluorescence Quantitation System (Promega Corporation, Madison, WI, USA). Finally, the samples were proportionally mixed according to the sequencing requirements for subsequent experiments.

2.5. 16RsRNA Sequence Amplification and Library Construction

To prevent excessive sequence length for sequencing while maintaining amplification heterogeneity, the primer set 515F/907R was selected for amplifying the bacterial 16S rDNA fragment from total DNA samples. Polymerase chain reaction (PCR) amplifications were conducted using the following gene-specific primers: 515F: 5′-GTGCCGCCAGCMGCCGG-3′, 907R: 5′-CCGTCAATTCMTTTRAGTTT-3′. PCR reactions were prepared in triplicate with a 20 μL reaction volume, and the DNA template concentration was 10 ng. To ensure accurate and reliable data, a lower cycle number was employed for amplification, aiming for uniform cycle numbers across all samples. After preliminary experimentation, 25 cycles were determined to be optimal. The PCR protocol consisted of an initial denaturation step at 95 °C for 2 min, followed by 25 cycles of denaturation at 90 °C for 30 s, annealing at 55 °C for 30 s, extension at 72 °C for 30 s, and a final extension at 72 °C for 5 min. The PCR products from the same sample were pooled and analyzed using 2% agarose gel electrophoresis. Gel extraction of PCR products was conducted using the AxyPrepDNA Gel Recovery Kit (Axygen Biosciences, Union City, CA, USA), followed by elution in Tris-HCl buffer and subsequent analysis via 2% agarose gel electrophoresis. For accurate quantification, the PCR products were quantified using the QuantiFluor™—ST Blue Fluorescence Quantitation System (Promega Corporation), based on the preliminary results obtained from gel electrophoresis. Subsequently, the samples were proportionally mixed according to the sequencing requirements for each sample. The mixed samples were subjected to high-throughput sequencing using the Illumina HiSeq 2500 platform (Shanghai BIOZERON Biotech., Shanghai, China). This sequencing strategy aimed to balance sequence length for effective sequencing while ensuring adequate representation of genetic diversity across the samples. All the sequences were deposited in the NCBI, and the serial number was PRJNA1059507.

2.6. Sequencing Data Processing and Analysis

The data extracted based on index sequences were stored in fastq format. Quality control filtering was performed on reads using barcode information to distinguish between different samples. After filtering, paired-end (PE) reads were analyzed to identify overlapping regions and subsequently merged into a single sequence. High-quality sequences were obtained for each sample by splitting based on barcode and primer sequences. Sequence directionality was corrected based on the orientation of the forward and reverse barcodes and primers, while chimeras were removed during this process. A quality control filter was applied to trim bases with a quality value below 20 at the end of the reads. Additionally, a sliding window of 10 base pairs was implemented to identify regions with an average quality value lower than 20, and bases beyond this window were trimmed. Reads shorter than 50 base pairs after quality control were filtered out. The PE reads were consolidated into a single sequence with a minimum overlap length of 10 base pairs. The maximum allowable mismatch rate in the overlapping region of merged sequences was set to 0.2, and sequences not meeting this criterion were removed. Samples were demultiplexed based on barcode and primer regions, with barcode mismatch set to zero and a maximum primer mismatch of 2. A combined de novo and reference approach using Usearch (V10) software and the Gold database was employed to remove chimeric sequences. This rigorous process ensured the generation of high-quality, non-chimeric sequences suitable for subsequent analysis.

For soil alpha diversity analysis, Chao1, Shannon, Simpson, and ACE indices in soil samples were determined using Mothur (version 1.35.1). This analysis was conducted using the Uclust algorithm (version 1.2.22q) to categorize species based on representative operational taxonomic unit (OTU) sequences at a 97% similarity threshold. Analysis of species composition and disparities within each soil sample was performed at both the phylum and genus levels. Additionally, redundancy analysis across different samples was conducted using the Canoco 5.0 software. Basic data analysis was conducted using Excel 2020, whereas the SPSS 26.0 software was employed for statistical analysis. Origin 2019 and the R programming language (3.6.3) were used for data visualization. This comprehensive array of tools and software enabled a thorough exploration and understanding of soil microbial communities and their diversity.

3. Results

3.1. Physicochemical Properties of the Soil

The physicochemical properties of soil varied significantly depending on land use (Table 1). Specifically, TN and AN reached their highest levels in industrial forests at 4.17 g/kg and 156.33 g/kg, respectively, whereas natural forests exhibited the lowest values of 1.67 g/kg for TN and 85.17 g/kg for AN. SOC reached its peak concentration in natural forests at 71.47 g/kg, contrasting with its lowest value in agricultural land, measuring 43.64 g/kg. TP content was 0.59 g/kg for farmland and 0.26 g/kg for natural forest. Lastly, TK exhibited the highest (9.76 g/kg) and lowest (6.25 g/kg) concentrations in natural forests.

Table 1. Soil physical and chemical properties of different land use types.

Soil Sample	TN (g/kg)	AN (mg/kg)	TP (g/kg)	AP (mg/kg)	TK (g/kg)	AK (mg/kg)	SOC (g/kg)
NF	1.672 ± 0.159 b	85.167 ± 5.09 b	0.264 ± 0.02 a	5.32 ± 0.67 b	6.25 ± 0.28 b	61.80 ± 4.32 a	71.47 ± 7.11 b
AF	4.166 ± 0.948 a	156.333 ± 21.61 a	0.479 ± 0.06 b	5.24 ± 0.85 b	9.76 ± 0.62 a	64.08 ± 20.46 a	64.23 ± 4.65 a
FL	3.155 ± 0.464 ab	135.333 ± 9.33 a	0.586 ± 0.32 b	30.76 ± 3.72 a	9.31 ± 0.334 a	86.42 ± 18.04 a	43.64 ± 3.77 a

Note: The data are means ± standard error (SE), different letters indicate significant levels ($p < 0.05$). a and b stand for Duncan's multiple-range test. TN (total nitrogen); AN (immediate nitrogen); TP (total phosphorus); AP (immediate phosphorus); TK (total potassium); AK (immediate potassium); SOC (organic carbon); NF (natural forest); AF (artificial forests); FL (farmland).

3.2. Microbial α-Diversity Analysis

Significant differences were observed among the soil bacterial α-diversity of different land uses (Table 2). The Chao1 index reached its peak at 4475.56 in agricultural land, whereas the lowest value (3775.49) was recorded in natural forest. Moreover, the highest Shannon's diversity index of 10.44 was observed in natural forests, whereas the lowest value (9.55) was detected in agricultural land. Similarly, Simpson's index exhibited the greatest diversity at 0.56 in natural forest, contrasting with its lowest value (0.027) in agricultural land.

Table 2. α Diversity of soil bacteria under different land uses.

Land Use Type	Chao1	Shannon	Simpson	Coverage
NF	4343.4 ± 14.939 a	10.2 ± 0.108 a	0.027 ± 0.001 b	0.974 ± 0.005 a
AF	3775.5 ± 159.176 b	10.41 ± 0.198 a	0.557 ± 0.011 a	0.974 ± 0.002 a
FL	4475.6 ± 116.811 a	9.6 ± 0.3555 b	0.193 ± 0.003 b	0.973 ± 0.003 a

Note: The data are means ± standard error (SE), different letters indicate significant levels ($p < 0.05$). a and b stand for Duncan's multiple-range test.

3.3. Structural Composition of Microbial Communities

The results of our high-throughput sequencing analyses yielded a total of 334,356 valid sequences from all of the examined soil samples, with an average length of 363.88 bp. Out of these sequences, 8260 sequences were categorized as OTUs and further classified into 41 phyla, 102 orders, 232 orders, 296 families, 523 genera, and 601 species (Figure A1). The number of OTUs varied among the different land use types: 5821 species in agricultural land, 5630 species in natural forests, and 5102 species in planted forests. There were a total of 3207 OTUs in the three sampling plots, with agricultural land, natural forests, and planted forests encompassing 1274, 1055, and 845 OTUs, respectively (Table A1).

As outlined in Table 3, significant differences were observed in the levels of bacterial phyla under different land uses, with Myxococcota, Acidobacteriota, Actinobacteriota, Chloroflexi, Gemmatimonadota, and Planctomycetota showing significant differences. In contrast, Bacteroidia, Proteobacteria, and Verrucomicrobiota were not significantly different (Table A2). The Acidobacteria phylum exhibited the highest abundance (31.2%) in natural forests and the lowest (14.6%) in agricultural land. The Actinobacteria phylum displayed the highest abundance (18.7%) in agricultural land and the lowest (11.8%) in natural forests, whereas the Gemmatimonadota phylum was the highest at 8.6% in agricultural land and the lowest at 3.8% in artificial forests (Figure 1).

Table 3. Spearman correlation analysis between soil physicochemical properties and microorganisms.

Phylum	Physical and Chemical Properties of the Soil						
	AK	AN	TN	TP	TK	AP	SOC
Myxococcota	0.2333	−0.569	−0.4833	−0.2333	**−0.6667 ***	0.35	−0.5167
Acidobacteriota	−0.2	0.2343	0.2333	−0.05	0.3833	−0.5333	0.4
Actinobacteriota	0.5	0.159	0.1333	0.45	−0.1333	**0.6833 ***	−0.1
Chloroflexi	0.4667	−0.1255	−0.0833	0.2333	−0.4	0.5167	−0.1
Bacteroidota	−0.3667	−0.0084	−0.15	−0.5333	−0.5833	−0.2833	−0.1353
Gemmatinnonadota	0.1333	**−0.7029 ***	**−0.6833 ***	−0.333	**−0.8333 ****	0.1167	**−0.75 ***
Planctomycetota	0.0833	0.3264	0.3533	−0.0333	0.35	−0.3833	0.5
Proteobacteria	−0.05	0.4268	0.3333	0.25	−0.0167	0.2667	0.3667
Verrucomicrobiota	0.1833	0.0418	0.0667	−0.0167	0.2167	0.0167	−0.1167
Chao1	0.35	−0.0335	−0.05	0.2833	−0.3833	0.4833	−0.3167
Shannon	0.4333	−0.1506	−0.1333	0.25	−0.4167	0.5	−0.4

Note: AK, AN, TN, TP, TK, AP, AK, AN, TN, TP, TK, AP, and SOC represent quick-acting potassium, quick-acting nitrogen, total nitrogen, total phosphorus, total potassium, quick-acting phosphorus, and organic carbon, respectively; * $p < 0.05$; ** $p < 0.01$. The bold indicates a significant correlation between soil physicochemical properties and soil bacterial phyla.

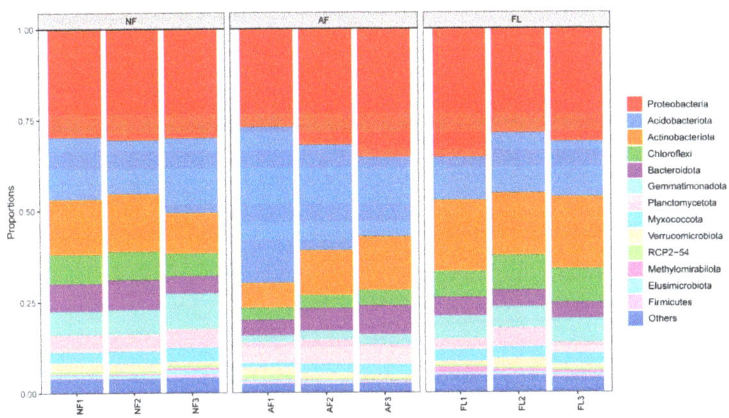

Figure 1. Bacterial community composition in different soil samples at the phylum level.

At the genus level (Figure 2), bacterial taxa with a relative abundance greater than 5% included *Sub-group 2_norank* (11.69%), *Gaiellales_norank* (6.6%), *Gemmatimonadaceae_uncultured* (6.21%), *Acidobacteriales_norank* (5.94%), and *Bradyrhizobium* (5.13%). Bacterial taxa with a relative abundance of less than 1% accounted for more than 30% of the three land use types.

3.4. Correlation between Soil Physical and Chemical Properties and the Relative Abundance of Microbial Communities

Redundancy analysis (RDA) of the soil biological communities of the three land use patterns at the genus level was conducted to preliminarily analyze the effects of soil physicochemical factors on soil microbial communities (Figure 3). The RDA results indicated that the two axes explained a total of 93.67% of the community changes, with the vertical axis explaining 8% of the community changes and the horizontal axis explaining 85.67% of the community changes. The composition of the bacterial community of natural forests was primarily influenced by AP and AK, whereas that of artificial forests was influenced by TK, AN, TN, and SOC, and the bacterial community composition of farmland soil was mainly influenced by TP, AP, and AK.

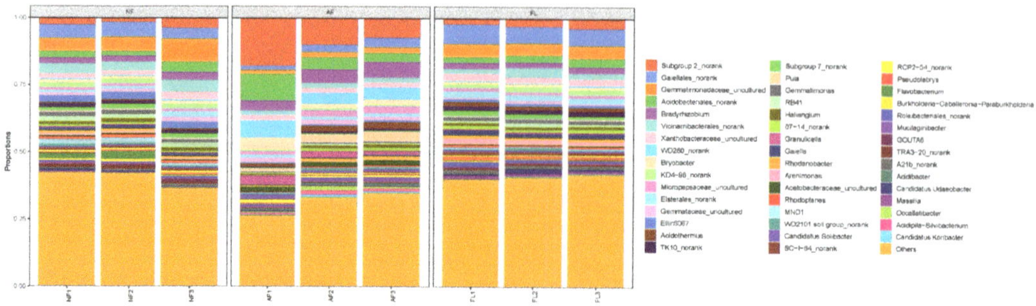

Figure 2. Bacterial community composition in different soil samples at the genus level.

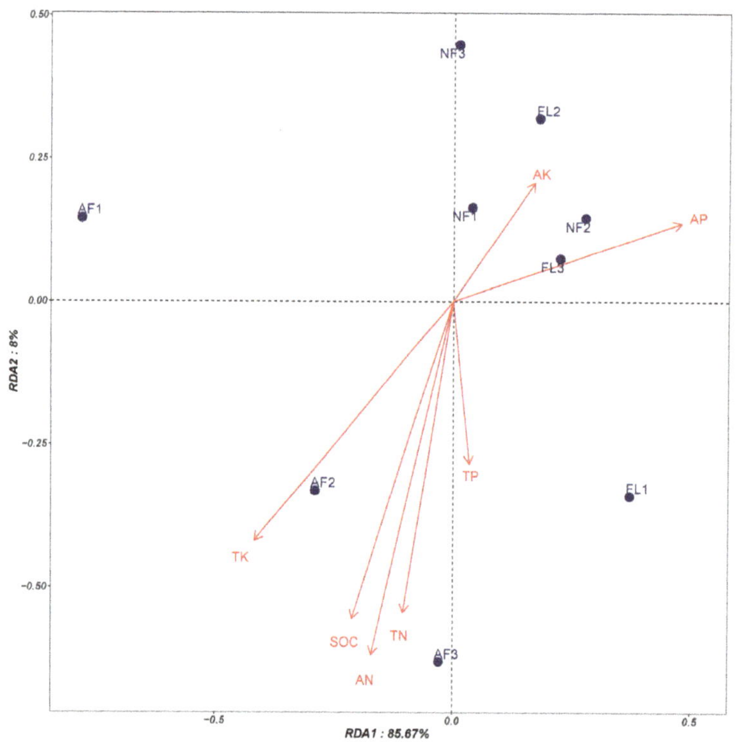

Figure 3. RDA of microbial community composition and soil physicochemical properties.

Spearman correlation analysis was conducted to reveal the relationship between the microbial community composition and the physical and chemical properties of the three soil types depending on land use (Table 3). The results revealed that the Chao1 index and Shannon index were not significantly correlated with physical and chemical properties. However, significant correlations existed between the dominant bacterial groups (Mucor, Actinobacteria, and Bacillus) and soil physicochemical properties. In contrast, the dominant bacterial phyla such as Aspergillus and Acidobacteria, did not exhibit significant correlations with any of the examined physicochemical properties. Specifically, the genus Mucor exhibited a significant negative correlation with total potassium content ($p < 0.05$), whereas the phylum Actinobacteria exhibited a significant positive correlation with quick-acting phosphorus content ($p < 0.05$). The phylum Bacillus displayed a significant negative correlation with quick-acting nitrogen content ($p < 0.05$), a highly significant negative correlation with total potassium content ($p < 0.01$), and a significant negative correlation with soil organic carbon content ($p < 0.05$).

4. Discussion

Soil microorganisms are crucial components of ecosystem dynamics and play a pivotal role in plant and soil nutrient cycling metabolism [16,17], with bacteria being among the most diverse and numerous microorganisms in soil [18]. Their multifaceted impact extends to plants, animals, and soils, influencing soil nutrient cycling, fertility maintenance, and enhancement, as well as plant growth and development [19].

This study revealed Proteobacteria, Acidobacteria, and Actinobacteria as dominant bacterial phyla in soils of different land use types. However, land use significantly altered the relative abundance of soil bacterial phyla, which was consistent with the results of Zou, Xiaoxiao and Cai, Guanxia [20–23]. Moreover, we summarized other studies in different regions and found that the dominant soil bacterial phyla were all the same in different land use types, but the structure and diversity of soil microbial communities changed significantly (Table A3). In this study, we observed substantial changes in the relative abundance of dominant phyla, with the highest relative abundance of Actinobacteria reaching 18.7% in agricultural land, whereas the lowest level of 14% was observed in natural forests. This variation may be attributed to manual straw crushing practices in farmland, leading to the highest relative abundance of Actinobacteria, which accelerates apoplastic material decomposition [24]. It may also be related to the potential ecological properties of actinomycetes, where bacteria of the phylum Actinomycetes play a key role in organic matter decomposition, facilitating the decomposition and cycling of organic matter. The highest relative abundance of Acidobacteria was identified in artificial forests at 31.2%, contrasting with the lowest in agricultural land at 14.6%. Although previous studies have suggested that Acidobacteria abundance is higher in nutrient-poor environments [25–27], the present findings align with Junjie Liu et al.'s observations in black soil, indicating a unique distribution pattern and a correlation with soil nutrient content [28]. It may also be related to the life habits of Acidobacteria, which are usually abundant in more acidic soils and are adapted to acidic environments. The highest abundance of Proteobacteria was observed in agricultural land at 31.8% and the lowest in natural forest at 30%. This may be linked to Proteobacteria preference for eutrophically rich soils [29], which is consistent with the relatively high nutrient content in agricultural land (nitrogen, phosphorus, and potassium). The adaptability of Proteobacteria, often parthenogenetic or aerobic bacteria, could also contribute to their higher abundance [30]. Additionally, artificial fertilization in farmland may have influenced the soil structure, enhancing the suitability of green leafy plants, known for their stress tolerance, in nutrient poor environments [31]. It is also possible that this may be related to the cultivation of soybean at the site, which is a nitrogen-fixing plant, and that some members of the phylum Ascomycota are involved in nitrogen fixation and nitrification-reduction processes, which have an important influence on the nitrogen cycle in the soil. However, the slightly higher relative abundance of the Gemmatimonadota phylum in agricultural land may be related to organochlorine pesticide use, as its abundance has been shown to increase with the application of organochlorine [32].

From Table 3, Actinobacteriota, Myxococcota, and Gemmatinnonadota were more strongly correlated with soil chemical properties. Among them, Actinobacteriota showed some degree of positive correlation on AK, AN, TN, TP, TK, AP, and SOC, especially on AP and SOC. This may be due to the fact that some members of Actinobacteriota may have the potential to decompose organic phosphorus. Organic phosphorus is an important pool of phosphorus in the soil, and its degradation helps to provide plant-available inorganic phosphorus. Thus, the positive role of Actinobacteriota may involve the production of organic phosphorus-degrading enzymes that promote efficient phosphorus utilization in the soil. On the other hand Actinobacteriota are usually known for decomposing organic matter. They may be involved in the decomposition of complex organic matter in the soil, producing secondary metabolites, and contributing to the soil carbon cycle. The positive correlation in terms of SOC may be related to the involvement of Actinobacteriota in organic matter degradation. They may decompose residues and organic matter and promote the decomposition and mineralization of organic matter in the soil. Myxococcota may be more sensitive to the utilization of fast-acting potassium. This may involve a decrease in the relative abundance of Myxococcota in areas enriched in fast-acting potassium, as their metabolic activities are affected by fast-acting potassium concentrations. The negative correlation of Myxococcota may be related to their specific functions in the nitrogen, phosphorus, and carbon cycles. Possible mechanisms include their utilization of certain nutrients in the soil or their sensitivity to specific environmental conditions. Gemmatinnonadota showed negative correlations for a wide range of soil nutrients (AK, AN, TN, TP, TK, AP), which may be due to its high sensitivity relative to the concentrations of these nutrients. It is possible that its metabolic activities are directly affected by changes in the concentrations of these nutrients. The negative correlation with TK and SOC may indicate that Gemmatinnonadota is highly sensitive to organic carbon levels in the soil. This group of bacteria may increase in relative abundance in environments with relatively low SOC in response to decreasing organic carbon levels.

At the genus level, a total of 523 microbial genera were detected in natural forests, artificial forests, and farmland under different land use modes over the long term, and the distribution patterns of these microbial genera were notably distinct. Subgroup 2_norank exhibited the highest relative abundance in plantation forest soils and belonged to the Acidobacteria phylum. This observation may be attributed to the predominance of eutrophic Acidobacteria in natural forests, where nutrient content is relatively high. It could also be linked to the unique distribution pattern of Acidobacteria in black soils [28], consistent with our study's findings. The relative abundance of Gaiellales_norank was higher in farmland than in natural and artificial forests, potentially connected to farmland tillage. Aerobic bacterium characteristics of Gaiellales_norank may be favored by crop tillage, increasing soil oxygenation [33].

Differences in soil utilization not only impact variations in microbial community structure but also influence the diversity of soil microbial communities [34]. In this study, significant differences in soil microbial community diversity were observed among different land use modes. The bacterial diversity of natural forests was higher than in other land use modes, consistent with the results of Jiao et al. [35]. The higher bacterial diversity in natural forests may be attributed to long-term management practices such as artificial nurturing and inter-logging, which influence the soil microbial community and promote or inhibit the growth of specific microorganisms, thus enhancing bacterial diversity [30]. Land use practices have substantial effects on soil chemical properties (e.g., TN, AN, SOC, TK, etc.), as well as on soil microbial diversity. Our RDA results revealed that soil SOC, TN, AN, and TK had a significant impact on soil bacterial diversity, aligning with previous studies indicating the strong influence of soil nutrients on the diversity and abundance of soil microbial communities [36]. Here, we observed that the lowest bacterial diversity index in farmland may be attributed to the lack of nutrients and organic matter in the soil due to prolonged cultivation, even with artificial fertilization. Additionally, factors such as the absence of ground cover, prolonged exposure of the soil surface to sunlight, and soil dryness may hinder the survival and reproduction of bacteria. The SOC content may also play a role, as studies suggest that the conversion of forests into farmland leads to a 25% to 50% decrease in SOC content in the soil [37]. Wang G et al. [38] found that lower SOC content may reduce the number of antagonistic bacteria in the soil, subsequently diminishing soil bacterial diversity. The results of this study suggest that changes in nutrient elements and chemical properties in soils of different land use types may directly affect the ecological niche and metabolism of microorganisms. Also, it may be related to vegetation types, which provide different root secretions and organic matter, which directly affect the type and number of microorganisms in the soil.

Related studies have indicated that soil microbial abundance increases after converting native vegetation to agricultural land [39], aligning with the findings of this study. The highest bacterial abundance was observed in agricultural land, possibly linked to anthropogenic disturbance. Lammel et al. noted a substantial increase in soil bacterial abundance from primary rainforests and savannahs to agricultural land in the southern part of the Amazon due to sustained anthropogenic disturbance [40]. Frequent anthropogenic disturbances may stimulate the enrichment of bacterial OTUs [41]. Natural forests exhibited relatively lower bacterial abundance, possibly due to the variable responses of different bacterial communities to changes in soil nutrient levels. Changes in the soil microenvironment, including nutrients, can also influence microbial diversity [42]. In this study, Spearman analysis indicated that SOC and TN had a lesser effect on soil diversity, contrary to previous studies suggesting a greater impact of soil nutrients on the diversity and abundance of soil microbial communities [43]. Seasonal factors must also be considered, as highlighted in the study by Ji Chuning et al., where pH and moisture content significantly affected the diversity of soil fungal communities during fungal community succession in reclaimed soils in a mining area [44]. Various factors, including seasonal variations, land use, and management practices, significantly impact bacterial abundance and diversity [45,46]. Sampling was conducted during the summer season, when plants naturally grow and differences in vegetation can cause changes in soil physicochemical properties, affecting soil microorganisms. Agricultural management practices and variations in climatic conditions during the summer may alter soil properties, subsequently influencing soil microorganisms. Although this study provides clear evidence that land use type significantly affects soil microbial community composition, there are still several uncertainties and limitations that must be addressed. The response of the microbial community to land use changes is time-dependent. However, this study only examined short-term responses. To better understand the impact of different land use types on the structure and diversity of soil microbial communities and provide a reliable basis, long-term and continuous monitoring across various land use types is essential.

5. Conclusions

Our findings demonstrated significant differences in soil bacterial diversity among different land use modes in the Black-billed Grouse Nature Reserve at the Central Station of Heilongjiang. Soil bacterial diversity in artificial forests was significantly higher than that in farmland, indicating a notable variation in microbial richness. The nutrient content of soil in artificial forests and farmland exhibited a significant increase compared to that in natural forests. Additionally, microbial abundance in farmland was significantly higher than that in artificial forests. Meanwhile, the dominant microbial taxa in natural forests, plantation forests, and farmlands were essentially the same, but their community structures differed. The primary environmental drivers for these variations might include TN, AN, SOC, and AP.

Author Contributions: The initial idea for this research was conceived by Y.L.; M.L. and L.M. performed the experiments, collected and analyzed the data, and wrote the manuscript; Y.L., R.X., H.H., X.L. and Q.C. performed some of the lab work; X.S. revised the manuscript. All authors have read and agreed to the published version of the manuscript.

Funding: This research was funded by the Open Grant for Key Laboratory of Sustainable Forest Ecosystem Management-Ministry of Education, School of Forestry, and Northeast Forestry University (KFJJ2023YB04), and Heilongjiang Provincial Ecological Environmental Protection Research Project (HST2022ST008).

Institutional Review Board Statement: Not applicable.

Informed Consent Statement: Not applicable.

Data Availability Statement: The data that support the findings of this study are available on request from the corresponding author.

Acknowledgments: We are grateful to Zhenzhu Zhao, employed in the Heilongjiang Zhongyangzhan Black-billed Capercaillie Nature Reserve, for allowing us access to the nature reserve, as well as to Fuyuan Chen, Xianda Li, and Qiu Guoliang for their assistance in sample processing.

Conflicts of Interest: The authors declare no conflict of interest.

Appendix A

Table A1. Summary of the high-throughput sequencing data.

Land Use Types	Sequences	Average Length	Coverage	OTU
NF	110,686	263.3	0.9775	5630
AF	109,092	263.2	0.9714	5102
FL	114,581	263.6	0.9776	5821

Table A2. ANOVA of dominant bacterial abundance at the phylum level (relative abundance > 1%).

Phylum	NF	AF	FL
Myxococcota	0.034 ± 0.003 a	0.022 ± 0.005 b	0.031 ± 0.006 ab
Acidobacteriota	0.175 ± 0.171 b	0.312 ± 0.06 a	0.146 ± 0.014 b
Actinobacteriota	0.14 ± 0.015 ab	0.113 ± 0.023 b	0.187 ± 0.008 a
Chloroflexi	0.073 ± 0.005 a	0.038 ± 0.003 b	0.086 ± 0.007 a
Bacteroidota	0.068 ± 0.01 a	0.061 ± 0.01 a	0.047 ± 0.002 a
Gemmatimonadota	0.077 ± 0.01 a	0.024 ± 0.002 b	0.063 ± 0.002 a
Planctomycetota	0.048 ± 0.002 ab	0.058 ± 0.004 b	0.037 ± 0.008 a
Proteobacteria	0.3 ± 0.002 a	0.315 ± 0.024 a	0.318 ± 0.019 a
Verrucomicrobiota	0.018 ± 0.004 a	0.015 ± 0.003 a	0.016 ± 0.005 a

Note: The data are means ± standard error (SE), different letters indicate significant levels ($p < 0.05$). a and b stand for Duncan's multiple-range test.

Table A3. Summary of data on dominant phyla and diversity in different regions.

Different Regions	Dominant Soil Bacterial Phylum (Top 4 Abundance)				α Diversity
The results of this study Natural forests	Proteobacteria	Acidobacteriota	Actinobacteriota	Chloroflexi	Agricultural land had the highest Chao1 index, while natural forests had the lowest values. In addition, the Shannon diversity index was highest in natural forests and lowest in agricultural lands.
Artificial forests	Proteobacteria	Acidobacteriota	Actinobacteriota	Chloroflexi	
Farmland	Proteobacteria	Acidobacteriota	Actinobacteriota	Chloroflexi	
Sichuan [23] Sandy wetland	Proteobacteria	Actinobacteriota	Acidobacteriota	Chloroflexi	Chao1 index showed muddy wetlands > sandy wetlands > natural forest land > farmland
Farmland	Proteobacteria	Actinobacteriota	Acidobacteriota	Chloroflexi	
mud wetland	Proteobacteria	Actinobacteriota	Acidobacteriota	Chloroflexi	
natural forest land	Proteobacteria	Actinobacteriota	Acidobacteriota	Chloroflexi	

Table A3. *Cont.*

Different Regions	Dominant Soil Bacterial Phylum (Top 4 Abundance)				α Diversity
Guizhou [21] bare soil	Proteobacteria	Acidobacteriota	Chloroflexi	Actinobacteriota	The Chao1 and α diversity of soil bacteria in grassland are both at high levels and significantly higher than those in other land use types. and significantly higher than other land use types.
Farmland	Proteobacteria	Acidobacteriota	Chloroflexi	Actinobacteriota	The abundance and alpha diversity of soil bacteria in grassland are both at high levels and significantly higher than those in other land use types. and significantly higher than other land use types.
Grassland	Proteobacteria	Acidobacteriota	Chloroflexi	Actinobacteriota	
Shurb	Proteobacteria	Acidobacteriota	Chloroflexi	Actinobacteriota	
GuangXi [20] Farmland	Actinobacteriota	Proteobacteria	Acidobacteriota	Chloroflexi	No significant differences in the richness and diversity indices were found among the three plots. Chao, Simpson, and Shannon diversity indices were the highest in farmland, followed by grassland and plantations.
Grass land	Actinobacteriota	Proteobacteria	Acidobacteriota	Chloroflexi	
Artificial forest	Actinobacteriota	Proteobacteria	Acidobacteriota	Chloroflexi	

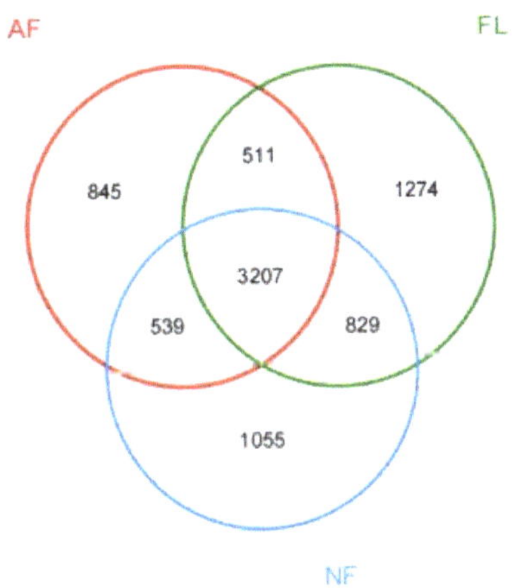

Figure A1. Venn diagram of the OTU distribution.

References

1. Sun, R.; Sun, B.H.; Gao, M.X.; Yang, X.Y.; Zhang, S.L. Changes of soil microbial characteristics under long-term different land use patterns on an anthropogenic loess soil. *J. Plant Nutr. Fertil.* **2015**, *21*, 655–663.
2. Mganga, K.Z.; Musimba, N.K.R.; Nyariki, D.M. Combining Sustainable Land Management Technologies to Combat Land Degradation and Improve Rural Livelihoods in Semi-arid Lands in Kenya. *Environ. Manag.* **2015**, *56*, 1538–1548. [CrossRef] [PubMed]
3. Li, L.; Jiang, J.L. Bacterial Community Analysis of Panax quinquefolium Rhizosphere Soil by High-throughput Sequencing Technology. *J. Chin. Med. Mater.* **2019**, *42*, 7–12.

4. Sui, X.; Zhang, R.T.; Yang, L.B.; Xu, N.; Chai, C.R.; Wang, J.F.; Fu, X.L.; Zhong, H.X.; Xing, J.H.; Zhang, Y.; et al. Effect of simulation nitrogen depositions on bacterial diversity of Deyeuxia angustifolia in wetland of Sanjiang Plain. *Pratacult. Sci.* **2016**, *33*, 589–598.
5. Yang, J.; Zhou, G.Y.; Tian, Y.Y.; Liu, Q.L.; Liu, C.F.; Yang, Q.; Zhou, C.J. Differential analysis of soil bacteria diversity in different mixed forests of Dalbergia odorifera. *J. Ecol.* **2015**, *35*, 8117–8127.
6. Paletto, A.; Meo, I.D.; Cantiani, P.; Guerrini, S.; Lagomarsino, A. Social perception of forest management: The case of the peri-urban forest of Monte Morello. *Forest@ J. Silvicult. For. Ecol.* **2018**, *15*, 29–39. [CrossRef]
7. He, Z.; Yuan, C.L.; Chen, P.R.; Rong, Z.Q.; Peng, T., Taimoor, H.F.; Wang, G.J.; Yan, W.D.; Wang, J. Soil Microbial Community Composition and Diversity Analysis under Different Land Use Patterns in Taojia River Basin. *Forests* **2023**, *14*, 1004. [CrossRef]
8. Yan, B.; Li, J.C.; Xiao, N.W.; Yue, Q.; Fu, G.; Liu, G.H.; Qiao, M.P. Urban-development-induced Changes in the Diversity and Composition of the Soil Bacterial Community in Beijing. *Sci. Rep.* **2016**, *6*, 38811. [CrossRef]
9. Zhou, J.Z.; He, Z.; Yang, Y.F.; Deng, Y.; Tringe, S.; Cohen, L.A. High-Throughput Metagenomic Technologies for Complex Microbial Community Analysis: Open and Closed Formats. *mBio* **2015**, *6*, e02288-14. [CrossRef]
10. Li, J.B.; Pokharel, P.; Liu, G.M.; Chen, J.L. Reclamation of desert land to different land-use types changes soil bacterial community composition in a desert-oasis ecotone. *Land Degrad. Dev.* **2020**, *32*, 1389–1399. [CrossRef]
11. Wu, Y.N.; Gao, W.F.; Zou, Y.; Dong, H.Y.; Yu, F.; Wang, H.; Cheng, Z. Effects of Land Use Conversion on the Soil Microbial Community Composition and Functionality in the Urban Wetlands of North-Eastern China. *Forests* **2022**, *13*, 1148. [CrossRef]
12. He, X.Y.; Su, Y.R.; Liang, Y.M.; Chen, X.B.; Zhu, H.H.; Wang, K.L. Land reclamation and short-term cultivation change soil microbial communities and bacterial metabolic profiles. *J. Sci. Food Agric.* **2012**, *92*, 1103–1111. [CrossRef] [PubMed]
13. Wang, X.H.; Yang, J.H.; Xie, X.F.; Chen, X.J.; Pu, L.J.; Zhang, X.W. Soil microbial succession with soil development since costal reclamation. *Catena* **2020**, *187*, 104393. [CrossRef]
14. Wang, D.W.; Wang, J.; Fu, G. Dynamic Assessment and Analysis of Grain Production Balance: A Study on Strategies for the Sustainable Development of Grain Security in the Western Region of Heilongjiang Province, China. *Hydra Sci. Tech.* **2007**, *6*, 9–10.
15. Cheng, Z.Q.; Yang, L.B.; Zhang, T.; Wang, W.H.; Yin, W.P.; Li, G.F.; Song, F.Q. Soil Microbial Functional Diversity Responses to Different Revegetation Types in Heilongjiang Zhongyangzhan Black-Billed Capercaillie Nature Reserve. *Res. Environ. Sci.* **2021**, *34*, 1177–1186.
16. Jia, P.L.; An, S.S.; Li, C.C.; Zeng, Q.C.; Wang, B.R.; Bai, X.J. Dynamics of Soil Nutrients and Their Ecological Stoichiometry Characteristics Under Different Longitudes in the East-West Forest Belt of Loess Plateau. *J. Soil Water Conserv.* **2020**, *34*, 315–321.
17. Yu, J.Y.; Huang, X.H.; Shuai, Z.B.; Xie, L.; Liu, Y.H.; Wang, C.R.; Chai, D.; Liu, D.; Zhou, X.Y.; Chen, Q. The community structure of bacteria and fungi in soils with root rot diseased garlic plants in Pengzhou, Sichuan Province. *Chin. J. Appl. Environ. Biol.* **2020**, *26*, 928–935.
18. Chen, X.R.; Nan, B.Z. Bacterial Diversity and Its Role in Agricultural Ecosystems. *Pratacult. Sci.* **2002**, *9*, 34–38.
19. Zhao, F.; Zhao, M.Z.; Wang, J.; Guan, L.; Pang, F.H. Studying the Microbial Community Structure and Diversity in the Rhizosphere of Strawberries Using High-Throughput Sequencing. *Soils* **2019**, *51*, 51–60.
20. Zou, X.X.; Yao, K.; Zeng, F.P.; Zhang, C.; Yao, Z.X.; Zhang, H. Diversity and Assembly of Bacteria Community in Lime Soil under Different Karst Land-Use Types. *Forests* **2023**, *14*, 672. [CrossRef]
21. Cai, G.X.; Xia, F.; Li, D.; Bao, Q.; Chen, B.; Zhao, M. A Simulation Study on Characteristics of Soil Microbial Community Structure under Different Land Uses. *Bull. Miner. Petrol. Geochem.* **2022**, *41*, 1023–1032.
22. Qin, H.; Li, C.X.; Ren, Q.S. Effects of different land use patterns on soil bacterial and fungal biodiversity in the hydro-fluctuation zone of the Three Gorges reservoir region. *Acta Ecol. Sin.* **2017**, *37*, 3494–3504.
23. Liu, K.H.; Xue, Y.Q.; Zhu, L.P.; Xu, F.; Zhu, Z.H.; Zhang, T.; Zhang, F.B. Effect of Different Land Use Types on the Diversity of Soil Bacterial Community in the Coastal Zone of Jialing River. *Acta Ecol. Sin.* **2022**, *43*, 1620–1629.
24. Foulon, J.; Zappelini, C.; Durand, C.; Valot, B.; Girardclos, O.; Blaudez, D.; Chalot, M. Environmental metabarcoding reveals contrasting microbial communities at two poplar phytomanagement sites. *Sci. Total Environ.* **2016**, *571*, 1230–1240. [CrossRef] [PubMed]
25. Zhang, Y.; Dong, S.K.; Gao, Q.Z.; Liu, S.L.; Zhou, H.K.; Ganjurjav, H.; Wang, X.X. Climate change and human activities altered the diversity and composition of soil microbial community in alpine grasslands of the Qinghai-Tibetan Plateau. *Sci. Total Environ.* **2016**, *562*, 353–363. [CrossRef] [PubMed]
26. Zhang, C.; Liu, G.B.; Sha, X.; Wang, G.L. Soil bacterial community dynamics reflect changes in plant community and soil properties during the secondary succession of abandoned farmland in the Loess Plateau. *Sci. Total Environ.* **2016**, *97*, 40–49. [CrossRef]
27. Ding, H.; Fang, Y.M.; Yang, Q.; Chen, X.; Yuan, F.Y.; Xu, H.; He, L.H.; Yan, J.; Chen, T.T.; Yu, C.J.; et al. Community characteristics of a mid-subtropical evergreen broad-leaved forest plot in the Wuyi Mountains, Fujian Province, Southeastern China. *Biodivers. Sci.* **2015**, *4*, 479–492. [CrossRef]
28. Liu, C.X.; Dong, Y.H.; Jiao, R.Z. Research Progress in Acidobacteria Diversity in Forest Soil. *World For. Res.* **2016**, *29*, 17–22.
29. Fierer, N.; Bradford, M.A.; Jackson, R.B. Toward an Ecological Classification of Soil Bacteria. *Bull. Ecol. Soc. Am.* **2007**, *88*, 1354–1364. [CrossRef]
30. Shen, L.; Li, Z.H.; Zeng, W.M.; Yu, R.L.; Wu, X.L.; Li, J.K.; Wang, L.K. Microbial Communities in Soils of Qingshuitang Industrial District in Zhuzhou. *Acta Ecol. Sin.* **2018**, *39*, 5151–5162.

31. Huang, S.; Pan, X.H.; Huang, Q.R.; Yu, X.C.; Li, C.X.; Zhang, W.J. Effect of Long-term Fertilization in Different Patterns on the Productivity and Soil Structure of Red Hilly Upland in Southern China. *Acta Agric. Univ. Jiangxiensis* **2012**, *34*, 403–408.
32. Costello, E.K.; Schmidt, S.K. Microbial diversity in alpine tundra wet meadow soil: Novel Chloroflexi from a cold, water-saturated environment. *Environ. Microbiol.* **2006**, *8*, 1471–1486. [CrossRef]
33. Cohen, J.K.; Zolti, A.; Harpaz, L.S.; Argaman, E.; Rabinovich, R.; Green, S.J.; Minz, D. Effects of tillage practices on soil microbiome and agricultural parameters. *Sci. Total Environ.* **2020**, *705*, 135791. [CrossRef]
34. Liu, Y.Q.; Ren, Y.R.; Liao, B.A.; Dong, Q.; Liu, M.S. Composition and structural characteristics of soil microbial communities in Yancheng typical coastal wetlands. *Acta Ecol. Sin.* **2023**, *43*, 2336–2347.
35. Jiao, S.; Chen, W.M.; Wang, Q.; Du, N.N.; Li, Q.P.; Wei, G.H. Soil microbiomes with distinct assemblies through vertical soil profiles drive the cycling of multiple nutrients in reforested ecosystems. *Curr. Stem Cell Res. Ther.* **2018**, *6*, 146. [CrossRef]
36. Zeng, J.; Shen, J.P.; Wang, J.T.; Hu, H.W.; Zhang, C.J.; Bai, R.; Zhang, L.W.; He, J.Z. Impacts of Projected Climate Warming and Wetting on Soil Microbial Communities in Alpine Grassland Ecosystems of the Tibetan Plateau. *Microb. Ecol.* **2018**, *75*, 1009–1023. [CrossRef] [PubMed]
37. Wu, L.Z.; Cai, Z.C. Effect of Agricultural Cultivation on Soil Organic Carbon in China. *J. Soil Water Conserv.* **2007**, *21*, 118–121+134.
38. Wang, G.Z.; Liu, Y.G.; Cui, M.; Zhou, Z.Y.; Zhang, Q.; Li, Y.J.; Ha, W.X.; Pang, D.B.; Luo, J.F.; Zhou, J.X. Effects of secondary succession on soil fungal and bacterial compositions and diversities in a karst area. *Plant Soil* **2021**, *475*, 1–12. [CrossRef]
39. George, P.B.L.; Lallias, D.; Creer, S.; Seaton, F.M.; Kenny, J.G.; Eccles, R.M.; Griffiths, R.I.; Lebron, I.; Emmett, B.A.; Robinson, D.A.; et al. Divergent national-scale trends of microbial and animal biodiversity revealed across diverse temperate soil ecosystems. *Nat. Commun.* **2019**, *10*, 1107. [CrossRef] [PubMed]
40. Lammel, D.R.; Nüsslein, K.; Veresoglou, C.E.P.C.S.D.; Rillig, M.C. Soil biota shift with land use change from pristine rainforest and Savannah (Cerrado) to agriculture in southern Amazonia. *Mol. Ecol.* **2021**, *30*, 4899–4912. [CrossRef]
41. Santillan, E.; Seshan, H.; Constancias, F.; Moses, D.I.D.; Wuertz, S. Frequency of disturbance alters diversity, function, and underlying assembly mechanisms of complex bacterial communities. *npj Biofilms Microbiomes* **2019**, *5*, 8. [CrossRef] [PubMed]
42. Zhang, R.H.; Li, R.; Zhang, L.L. Soil nutrient variability mediates the effects of erosion on soil microbial communities: Results from a modified topsoil removal method in an agricultural field in Yunnan plateau, China. *Environ. Sci. Pollut. Res. Int.* **2021**, *29*, 3659–3671. [CrossRef] [PubMed]
43. Yan, N.; Marschner, P.; Cao, W.H.; Zuo, C.Q.; Qin, W. Influence of salinity and water content on soil microorganisms. *Int. Soil Water Conserv. Res.* **2015**, *3*, 316–323. [CrossRef]
44. Liu, J.K.; Engel, A.B.; Wang, Y.; Zhang, G.F.; Zhang, Z.M.; Zhang, M.X. Multi-scale analysis of hydrological connectivity and plant response in the Yellow River Delta. *Sci. Total Environ.* **2020**, *702*, 134889. [CrossRef]
45. Daniel, L.F.; Eugen, U.; Mihaela, B.A.; Feodor, F.; Cristina, M.E. Diversity of soil bacteria as Indicator of Soil Pollution in Moldavia Region, Romania. *Environ. Eng. Manag. J.* **2017**, *16*, 879–889.
46. Steenwerth, K.L. Links between Soil Microbial Communities and Transformations of Soil Carbon and Nitrogen along a Gradient in Land-Use History and Soil Disturbance. Ph.D. Thesis, University of California, Davis, CA, USA, 2003.

life

MDPI

Article

Resilience of *Chlorella vulgaris* to Simulated Atmospheric Gas Compositions of Mars, Jupiter, and Titan

Ariela Likai †, Aikaterini Papazi † and Kiriakos Kotzabasis *

Department of Biology, University of Crete, Voutes University Campus, GR-70013 Heraklion, Crete, Greece
* Correspondence: kotzab@uoc.gr
† These authors contributed equally to this work.

Abstract: This study investigates the resilience of the unicellular green microalga *Chlorella vulgaris* to extreme atmospheric conditions simulating those of Mars, Jupiter, and Titan. Using Earth as a control, experiments were conducted under autotrophic and mixotrophic conditions to evaluate the organism's photosynthetic efficiency, oxygen production, and biomass growth over 2, 5, and 12 days. Photosynthetic performance was analyzed through chlorophyll a fluorescence induction (JIP-test), metabolic activity via gas chromatography, and biomass accumulation measurements. Despite the extreme atmospheric compositions—ranging from the CO_2-rich, low-pressure Martian atmosphere to the anoxic atmospheres of Jupiter and Titan—*C. vulgaris* demonstrated resilience and a functional photosynthetic apparatus, maintaining growth and oxygen production. Notably, the Martian atmosphere enhanced photosynthetic performance, with fluorescence curves and Fv/Fm ratios surpassing Earth-like conditions, likely due to elevated CO_2 and low pressure. Under mixotrophic conditions, the addition of glucose further enhanced metabolic activity and biomass growth across all atmospheres. These findings highlight the potential of *C. vulgaris* for bioregenerative life support systems, enabling oxygen production, CO_2 sequestration, and resource cultivation in extraterrestrial habitats. The study showcases the organism's adaptability to extreme environments, with implications for astrobiology, space exploration, and sustainable extraterrestrial ecosystems. These findings expand habitability criteria and explore extremophiles' potential to support life beyond Earth.

Keywords: *Chlorella vulgaris*; photosynthesis; extraterrestrial environments; Martian atmosphere; Jovian atmosphere; Titan atmosphere; metabolic plasticity

1. Introduction

Earth's atmosphere is a gaseous layer surrounding the planet, held in place by gravity, and extends to the Kármán line at approximately 100 km above sea level, where space is conventionally considered to begin. Atmospheric drag on spacecraft becomes noticeable below altitudes of about 150 km. The boundary between the atmosphere and space is not clearly defined; it gradually thins out and dissipates into space. Earth's atmosphere protects life by absorbing ultraviolet solar radiation, retaining surface heat through the greenhouse effect, and moderating temperature fluctuations between day and night. The composition of dry air is primarily 78.08% nitrogen (N_2), 20.95% oxygen (O_2), 0.93% argon, 0.04% carbon dioxide (CO_2), and traces of other gases [1]. When comparing the atmospheric compositions of some planets in our solar system (excluding their densities), like the atmospheres of Jupiter (90% H_2 and 10% He), Titan (a moon of Saturn—98% N_2 and 2% CH_4), and Mars (approximately 100% CO_2 at a pressure of <10 mbar), significant differences emerge relative to Earth's atmosphere.

https://doi.org/10.3390/life15010117

The selection of Mars, Titan, and Jupiter for this study reflects their distinct atmospheric compositions and relevance in astrobiological research. Mars and Titan are among the few bodies in the solar system with solid surfaces, making them of particular interest for studying potential habitability. Mars offers a CO_2-rich atmosphere and low pressure, while Titan has a nitrogen-dominated atmosphere with trace hydrocarbons. These unique environments provide valuable test conditions for the adaptability of terrestrial organisms. Jupiter was included as a representative of the gas giant planets, with its hydrogen-helium-dominated atmosphere serving as a model for understanding how photosynthetic organisms might behave in such extreme, anoxic environments.

Microalgae exhibit remarkable adaptability to various extreme environments [2–4]. The ability of microalgae to thrive in extreme environmental conditions has been extensively studied [2–5]. The microalga *Chlorella vulgaris* is notable for its exceptional tolerance to extreme conditions, including high CO_2 concentrations. This feature is crucial for environmental and astrobiological applications. Studies have shown that *C. vulgaris* can thrive in atmospheres with up to 60% CO_2 without significant stress, performing photosynthesis and producing oxygen simultaneously [2]. This resilience makes it suitable for carbon capture applications, such as near power plants. High CO_2 concentrations also enhance *Chlorella's* photosynthetic activity, boosting biomass and O_2 production while protecting its photosynthetic mechanism from photoinhibition [2].

Moreover, Wang et al. [3] have shown that *Chlorella* withstand space-like conditions. When exposed to high altitudes using scientific weather balloons, *Chlorella* cells experienced low temperatures, intense UV radiation, and reduced atmospheric pressure. While these harsh conditions caused significant physiological damage—such as DNA damage, reduced cell viability, and substantial mortality—the microalga was able to survive. Transcriptomic analyses revealed adaptive responses, including the expression of 3292 genes associated with heat shock proteins, antioxidant enzymes, DNA repair systems, and ribosomal proteins [3]. Notably, this study focused on survival rather than active growth or photosynthesis at stratospheric altitudes.

Additionally, under drought conditions, *Chlorella* develops protective mechanisms to mitigate stress effects, such as rapid recovery of photosynthetic activity, increased chlorophyll content, and accumulation of osmotic regulators like sucrose and trehalose [4]. Furthermore, *C. vulgaris* has been investigated for its potential role in bioregenerative life support systems (BLSS) due to its ability to produce essential biomolecules such as proteins, lipids, vitamins, and minerals [6].

Zerveas et al. [5] demonstrated that microalga *Scenedesmus obliquus*, in closed systems under anoxic conditions, can alter the atmosphere by enriching it with oxygen (O_2) through photosynthesis using solar energy. Specifically, when exposed to atmospheres containing 100% nitrogen (N_2) or 100% helium (He), devoid of CO_2 and O_2, the microalga seems to metabolize organic reserves, such as starch, through respiration to produce CO_2. This CO_2 is then used in photosynthesis to generate O_2. Part of the produced O_2 enriches the atmosphere, while the rest supports metabolic processes like respiration and biomass production. The recycling of products and substrates between photosynthesis and respiration contributes to an oxygen-enriched atmosphere and the microalgae's survival.

Microalgae are photosynthetic organisms capable of harnessing solar energy and absorbing large amounts of carbon dioxide, thus contributing to the production of atmospheric oxygen on Earth. *Chlorella vulgaris* has served as an essential model organism due to its simple structure, rapid growth under autotrophic, mixotrophic, and heterotrophic conditions, and high photosynthetic efficiency [7]. *C. vulgaris* consists of microscopic spherical cells with diameters ranging from 2 to 10 μm. The reproduction of *C. vulgaris* is asexual and occurs through a process called autosporulation. During autosporulation, the parent cell produces daughter cells within its cell wall before releasing them into the environment [7,8].

The aim of this study is to investigate the resilience and metabolic functionality of the unicellular eukaryotic microalga *C. vulgaris* under conditions simulating the atmospheres of Mars, Jupiter, and Titan. Specifically, it examines whether this microalga can survive, photosynthesize, and grow in extreme extraterrestrial atmospheres significantly different from Earth's. The findings will deepen our understanding of the microalga's immediate adaptive mechanisms in extreme conditions, highlighting its astrobiological and astrobiotechnological potential.

2. Materials and Methods

2.1. The Microalga *Chlorella vulgaris*—Culture Conditions and Experimental Procedures

A wild type strain of the unicellular green alga *Chlorella vulgaris* (SAG Number 211-11b) (Beyerinck, 1890) was cultured using Bold Basal Medium (BBM) as the nutrient medium [9]. Initial *C. vulgaris* cultures were prepared in 250 mL glass tubes shaped like saxophones (dimensions: 5 cm × 50 cm). The "saxophones" had an open top covered with hydrophobic cotton, and an air pump that provided continuous airflow (~50 L/h) through a tube at the bottom, ensuring optimal aeration and mixing to prevent sedimentation and enhance nutrient exchange (Supplementary Figure S1). These cultures were maintained under constant light intensity (about 50 μmol m^{-2} s^{-1}) and room temperature (25 °C) for one week, serving as mother cultures for subsequent experimental procedures.

Both autotrophic and mixotrophic BBM media were prepared for experimental procedures. The autotrophic medium was identical to the original BBM, while the mixotrophic medium included an additional 5 g/L glucose to provide an external organic carbon source. All experiments were conducted in hermetically sealed 125 mL glass vials with septa (dimensions: 5 cm × 9.5 cm), simulating the atmospheric conditions of Earth (atmospheric air—control), Jupiter (90% H_2 and 10% He), Titan (98% N_2 and 2% CH_4), and Mars (about 100% CO_2). Each vial contained 50 mL of the medium and 75 mL of the gaseous phase. The initial cell concentration was 1 μL PCV (packed cell volume)/mL. In all cases except the Mars simulation, the atmospheric pressure was approximately 1000 mbar. For Mars, the pressure was much lower (<10 mbar), achieved using a SpeedVac Vacuum Concentrator [10]. To establish a 100% CO_2 atmosphere, a compressed CO_2 gas cylinder was used. Using two sterilized needles (one for gas inflow and the other for outflow), the bottles were filled while maintaining a continuous gas flow to degas the medium and completely replace the air atmosphere. It took approximately 2 min to completely replace the bottle's atmosphere with CO_2. The atmospheric composition inside the bottle was verified using gas chromatography with thermal conductivity detection (GC-TCD). To achieve extremely low pressures nearing a vacuum (<10 mbar), a Heto freeze-dryer system (Heto, Maxi Dry Lyo model, Allerod, Denmark) equipped with a vacuum pump was employed. This system is capable of simulating pressures close to a vacuum (<10 mbar). After establishing the desired atmospheric composition in each closed culture system (100% CO_2), the septum-sealed bottles were placed inside the SpeedVac chamber. During the initiation of vacuum incubation (<10 mbar), a syringe was inserted into the septum of each bottle. This allowed the pressure of the atmosphere inside the bottles to adjust to match the atmospheric pressure of the SpeedVac chamber (<10 mbar) without altering the composition of the atmosphere within the bottles (100% CO_2). Under these conditions, the liquid medium began to evaporate slowly and gradually under low pressure. To address this issue, we stopped the vacuum after approximately five days of incubation, replenished the evaporated water with distilled water, repeated the gas phase formation process (100% CO_2) as described earlier, and resumed incubation until the 12th day.

For Jupiter's (90% H_2 and 10% He) and Titan's (98% N_2 and 2% CH_4) atmospheres, cultures were initially prepared using 100% H_2 and 100% N_2, respectively. To create Jupiter's atmosphere, 10% of the 100% H_2 atmosphere was removed using a gas-tight syringe, and 10% He was added. Similarly, for Titan's atmosphere, 2% of the 100% N_2 atmosphere was removed and 2% CH_4 was introduced. The atmospheric composition inside the bottle was verified using gas chromatography with thermal conductivity detection (GC-TCD).

We note that this study aimed to replicate the gas compositions and pressures of these extraterrestrial environments while maintaining laboratory temperatures (25 °C) conducive to sustaining *C. vulgaris*, as the goal was to assess the resilience of the organism's photosynthetic and metabolic activity under these atmospheric conditions.

2.2. Fluorescence Induction Measurement (JIP-Test)

Fluorescence induction was measured using the Plant Efficiency Analyzer (Handy PEA, Hansatech Instruments, King's Lynn, Norfolk, UK). The JIP-Test is a widely used technique in plant physiology to evaluate the efficiency of the photosynthetic apparatus, particularly of Photosystem II (PSII).

During photosynthesis, only part of the energy absorbed by pigments is used for photochemical processes, while the rest is emitted as heat or fluorescence. Fluorescence induction in photosynthetic organisms, first observed by Kautsky and Hirsch [11], exhibits two phases: rapid and slow [12]. Today, analyzing the fluorescence induction curve, particularly the rapid phase, is a tool for investigating the structure and function of the photosynthetic apparatus.

Rapid fluorescence phase measurements were recorded at 10 μs intervals over one second. Fluorescence was measured at a 12-bit resolution, with excitation via three red-light LEDs (intensity 3000 μmol m^{-2} s^{-1}). For fluorescence induction measurements, we placed the flat bottom of the small culture bottle (Ø 5 cm) directly on the PEA sensor using a specialized adapter that ensured complete darkness for at least 5 min, allowing the PSII reaction centers to fully open before exposure to saturated light. Under these conditions, and without opening the culture bottles, we conducted measurements of the culture in ambient conditions using the JIP-test [13]. Data analysis was conducted using Biolyser HP 4.0 software.

Key parameters studied included the Fv/Fm ratio (indicative of the maximum quantum efficiency of photosystem II and overall photosynthetic efficiency), Sm (a measure of the electron transport reservoir size), dV/dt_0 (electron transfer speed), N (number of Q_A reduction cycles required to close all reaction centers), TR_0/RC (energy trapped per reaction center), ET_o/RC (electron flow per reaction center), RC/CS_0 (density of active reaction centers per cross-sectional area), ABS/RC (functional antenna size per reaction center), DI_0/RC (non-photochemical energy dissipation per active center), and PI(abs) (a performance index combining the efficiency of energy absorption, trapping, and electron transport). Additionally, it is important to note that chlorophyll a fluorescence, particularly the steady-state fluorescence compared to the transient peak, provides complementary information to O_2 evolution. This relationship underscores that the fluorescence measurements reflect overall photosynthetic activity, not solely the status of photosystems.

2.3. Gas Chromatography—Thermal Conductivity Detection (GC-TCD) for Photosynthetic Activity

For qualitative and quantitative gas analysis, particularly oxygen, in closed culture systems, gas chromatography using thermal conductivity detection (GC-TCD) (Shimadzu GC 2010 Plus, Kyoto, Japan) was employed, with argon as the carrier gas at 5 bar pressure. A 250 μL gas sample was injected using a gas-tight precision syringe. Gases (O_2 and N_2) were separated based on thermal conductivity differences. The thermal conductivities of argon, nitrogen, and oxygen were 0.0001772 W/cmK, 0.0002598 W/cmK, and 0.0002674 W/cmK, respectively. A long capillary column (Ø 5 Å) was used, with detector temperature maintained at 200 °C, oven at 120 °C, and injection point at 180 °C. Quantitative gas analysis was performed using reference curves derived from known gas concentrations. According to the GC-TCD measurements for photosynthetic activity estimation, we recorded O_2 production during the first 2 days of incubation time (GC-TCD measurements at 0 and 2 days). Additionally, we know the biomass in the closed bottles (in μL PCV/mL) and the photosynthetic rate, expressed as $mLO_2 \cdot mL\ PCV^{-1} \cdot day^{-1}$.

2.4. Measurement of Packed Cell Volume (PCV)

To measure the packed cell volume of individual cultures, 1000 μL samples were centrifuged in calibrated capillary tubes for 5 min at 1500 g to sediment the cells [10]. Cellular concentration was expressed as packed cell volume (PCV) per mL of culture. This method assessed the growth rate of microalgal cultures under different conditions.

2.5. Data Analysis

All the experiments were conducted with a minimum of three repetitions to ensure reliability. Error bars on the diagrams represent the standard error of the mean (SEM), providing a measure of variability and precision in the sample means.

3. Results

3.1. Resilience of the Autotrophic Culture Chlorella vulgaris in the Atmospheres of Mars, Jupiter and Titan

This experiment examined the ability of the eucaryotic unicellular green microalga *Chlorella vulgaris* to survive and maintain its photosynthetic activity under conditions simulating the atmospheres of Jupiter, Titan (a moon of Saturn), and Mars. To assess whether the microalga could survive in these extraterrestrial environments, several parameters were monitored over time. These included the functional performance of the photosynthetic apparatus through chlorophyll a fluorescence induction measurements (JIP-test), its ability to metabolize and produce oxygen via photosynthesis using gas chromatography with thermal conductivity detection (GC-TCD), and its capacity to multiply assessed by measuring the growth of the cultures, expressed in μL PCV/mL. The Earth's atmosphere was used as a control, and the experiments were conducted under both autotrophic and mixotrophic conditions. Measurements were taken after incubating the cultures in the different atmospheres for 2, 5, and 12 days.

The atmospheres of the other celestial bodies differ significantly from that of Earth. The compositions used were as follows: Earth's atmosphere consisted of 78% N_2, 21% O_2, and 0.04% CO_2; Jupiter's atmosphere contained 90% H_2 and 10% He; Titan's atmosphere consisted of 98% N_2 and 2% CH_4; and Mars' atmosphere was composed of approximately 100% CO_2 at a pressure of <10 mbar. In all experimental setups, fluorescence induction curves were examined to provide fundamental insights into photosynthetic efficiency under autotrophic conditions in different planetary environments.

Fluorescence induction measurements reflect the functionality of the photosynthetic apparatus. On day 2, the fluorescence curves were nearly identical across all atmospheres and retained the characteristic shape of the OJIP curve observed in the control (Earth). This indicated proper functioning of the photosynthetic mechanism and efficient electron flow despite the extreme atmospheric conditions, suggesting that *C. vulgaris* under autotrophic conditions was capable of surviving and maintaining photosynthetic activity under these anoxic extreme conditions (Figure 1).

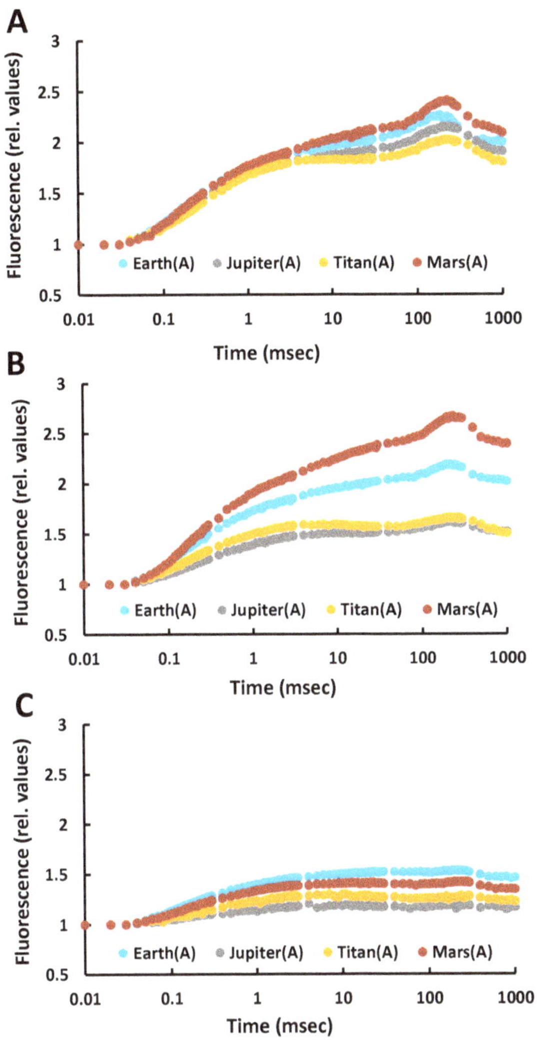

Figure 1. Fluorescence induction curves for all treatments with autotrophic cultures of the microalga *Chlorella vulgaris* under different simulated atmospheres (Jupiter, Titan, Mars) compared to the control treatment (Earth atmosphere) after incubation under these conditions for (**A**) 2 days, (**B**) 5 days, and (**C**) 12 days.

By day 5, the fluorescence curves began to diverge between atmospheres. The Earth's curve showed a slight decline but retained its characteristic shape, whereas those for Jupiter and Titan displayed noticeable drops, indicating stress on the cells. However, the curves still maintained the characteristic OJIP profile. On Mars, the fluorescence curve exceeded that of Earth, likely due to the CO_2 rich atmosphere (100%) combined with low atmospheric pressure (<10 mbar), which appeared to enhance photosynthesis (Figure 1).

By day 12, the curves had decreased in amplitude and steepness across all atmospheres, but still indicated photosynthetic activity. The overall decline, including in Earth conditions, was attributed to nutrient depletion in the closed system rather than atmospheric composition. These findings suggest that *C. vulgaris* can survive and photosynthesize in the atmospheres of Jupiter, Titan, and Mars under autotrophic conditions without prior adaptation. However, it is important to note that these results are based on controlled simulations and do not account for the absence of liquid water in such environments.

Further investigations examined structural and functional changes in the photosynthetic mechanism across conditions using the JIP-test. Figure 2 presents normalized parameter values relative to the control (Earth). Key parameters studied included the Fv/Fm ratio (indicative of photosynthetic efficiency), Sm (a measure of the electron transport reservoir), dV/dt_0 (electron transfer speed), N (number of Q_A reduction cycles), TR_0/RC (energy trapped per reaction center), ET_o/RC (electron flow per reaction center), RC/CS_0 (active reaction center density), ABS/RC (functional antenna size), DI_0/RC (non-photochemical energy dissipation per active center), and PI(abs) (photosynthetic efficiency).

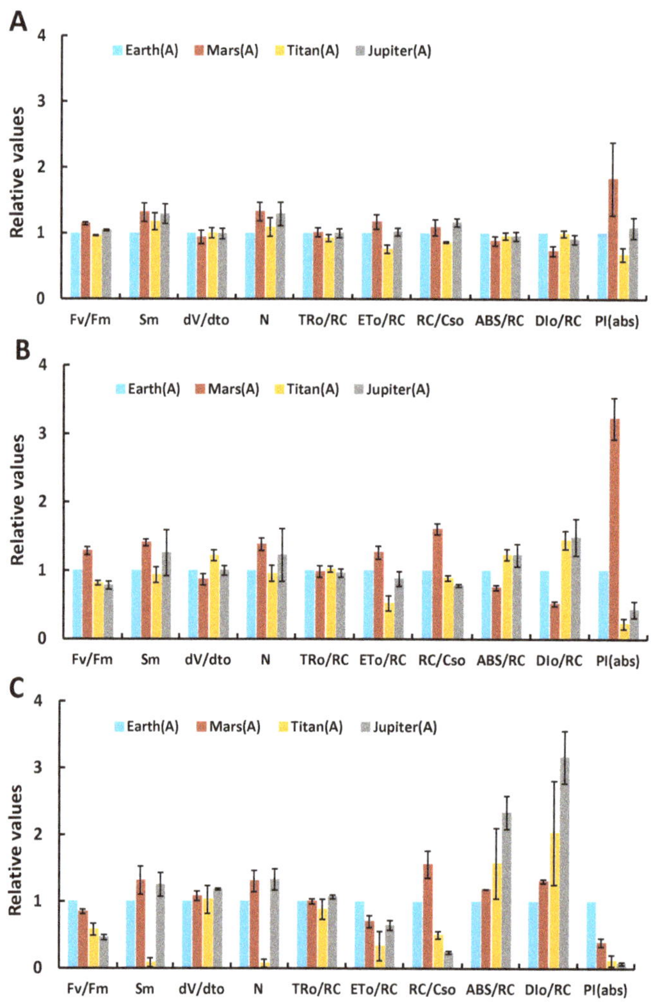

Figure 2. Differentiation of selected parameters from the JIP test (highlighting the structure and function of the photosynthetic mechanism) for treatments under simulated Jupiter, Titan, and Mars atmospheres compared to the control treatment (Earth atmosphere) under autotrophic conditions, after incubation under these conditions for (**A**) 2 days, (**B**) 5 days, and (**C**) 12 days.

On day 2, the Fv/Fm ratio on Mars was higher than that on Earth, indicating increased photosynthetic efficiency, likely due to the high CO_2 concentration [10]. In Titan's atmosphere, Fv/Fm was slightly lower than Earth's, while in Jupiter's atmosphere, it was slightly elevated. Sm values were higher for all three extraterrestrial atmospheres compared to Earth, reflecting a larger electron transport reservoir (the number of available electron carriers between the reaction centers of photosystem II and the plastoquinone pool). The dV/dt_0 values were similar across all atmospheres, suggesting functional and healthy photosystems. The RC/CS_0 and ABS/RC values were also comparable to Earth's, indicating no stress. PI(abs), a sensitive parameter reflecting PSII functionality, showed comparable values between Earth, Jupiter, and Titan, with Mars exhibiting significantly higher values, indicating robust photosynthetic activity even after 48 h of incubation in extreme conditions.

On day 5 (120 h of incubation), parameter values across the planets showed no significant deviations from Earth's normalized values, except for PI(abs) and DI_0/RC. Jupiter and Titan cultures demonstrated slight stress in their photosynthetic apparatus. Stress responses included decreased active center density (RC/CS_0), increased antenna size (ABS/RC), higher non-photochemical energy dissipation (DI_0/RC), and reduced Fv/Fm and PI(abs). In contrast, the Mars atmosphere showed enhanced photosynthetic performance, expressed as increased Fv/Fm and PI(abs). These changes suggest that the microalgae remained capable of photosynthesis despite moderate structural and functional adjustments to the PSII.

By day 12 (288 h of incubation), significant gradients emerged between Earth's parameters and those of other planetary atmospheres, reflecting stress without PSII collapse. All cultures, even those in Earth-like conditions, exhibited stress, attributed to nutrient depletion in the closed system. These stressors influenced PSII efficiency, electron transport, and energy management. However, the photosynthetic apparatus remained functional across all conditions.

To confirm metabolic functionality, net photosynthesis was measured (expressed as mL $O_2 \cdot mL\ PCV^{-1} \cdot day^{-1}$) using GC-TCD, except under Mars conditions where low pressure prevented measurement. The results showed oxygen production in all tested atmospheres. Earth's atmosphere supported the highest photosynthetic activity, followed by Titan and then Jupiter. This confirmed active metabolism and oxygen production by the microalgae despite adverse atmospheric conditions (Figure 3).

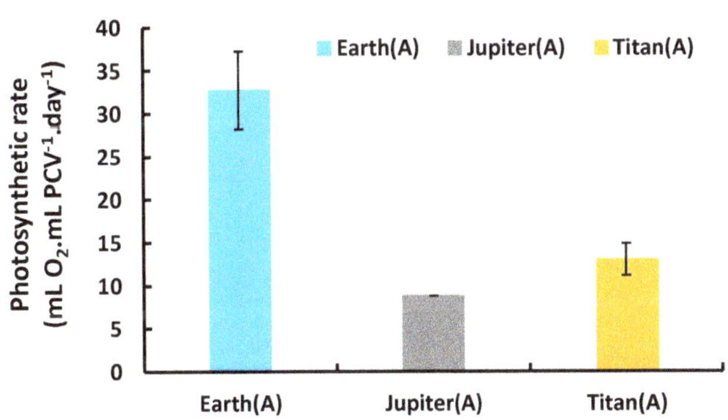

Figure 3. Photosynthetic O_2 production rate of *Chlorella vulgaris* under autotrophic conditions in different atmospheres.

Lastly, we assessed the ability of *C. vulgaris* to multiply under these extreme conditions by monitoring biomass changes over time, expressed as μL PCV/mL. All cultures started at 1 μL PCV/mL and exhibited biomass growth over time, indicating active proliferation in all atmospheric conditions (Figure 4). The distinct biomass production trends observed under autotrophic conditions in the reducing atmospheres of Jupiter and Titan compared to the oxidizing conditions of Earth and Mars suggest potential shifts in metabolic pathways. In the methane-rich atmosphere of Titan, Chlorella vulgaris may partially rely on alternative metabolic processes, such as fermentative or anoxygenic pathways, to build biomass, which could explain the surge followed by a plateau. This interpretation is supported by the inverse relationship between biomass production (Figure 4) and oxygen production (Figure 3) under these conditions.

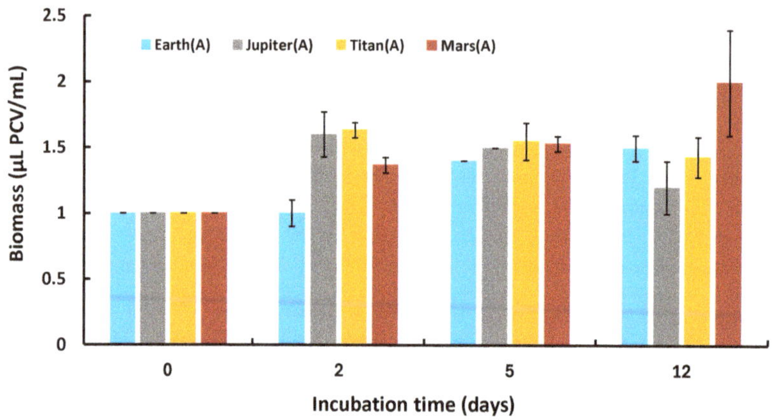

Figure 4. Biomass differentiation kinetics of *Chlorella vulgaris* for all treatments under autotrophic conditions.

These findings suggest the potential capability of microalgae to survive, photosynthesize, and reproduce under the simulated extreme atmospheres of Jupiter, Titan, and Mars. However, it is important to note that these results are based on controlled simulations and do not account for the absence of liquid water in such environments.

3.2. Resilience of Mixotrophic Chlorella vulgaris Cultures in the Atmospheres of Mars, Jupiter and Titan

We conducted experiments under mixotrophic conditions by supplementing inorganic nutrients with glucose (5g/L) to examine the behavior of mixotrophic cultures in various planetary atmospheres. The fluorescence induction curves of *C. vulgaris* under mixotrophic conditions in four different atmospheres (Earth, Jupiter, Titan, Mars) and after different incubation periods (2, 5, and 12 days) were comparable to or even higher than those observed under autotrophic conditions (Figure 5). This indicates that the microalga retains a highly active photosynthetic system. The fluorescence induction curve under Earth's atmosphere was sharp and high, nearly overlapping with the curve observed under Jupiter's atmosphere. The Mars curve was the highest among all atmospheres, while Titan's curve was the lowest. These findings suggest a healthy and active photosynthetic mechanism in the microalga (Figure 5A).

By Day 5, the fluorescence curves remained almost unchanged from Day 2. The Mars atmosphere continued to exhibit the highest fluorescence curve, while the curves for Earth and Titan remained at similar levels, and Jupiter's curve showed a slight decline (Figure 5B). By Day 12, the fluorescence curves under mixotrophic conditions were similar to those observed on Day 5, albeit slightly reduced (Figure 5C). These results suggest that the microalgae can maintain its photosynthetic capacity over longer periods when utilizing both light and an external organic carbon source, regardless of the incubation atmosphere.

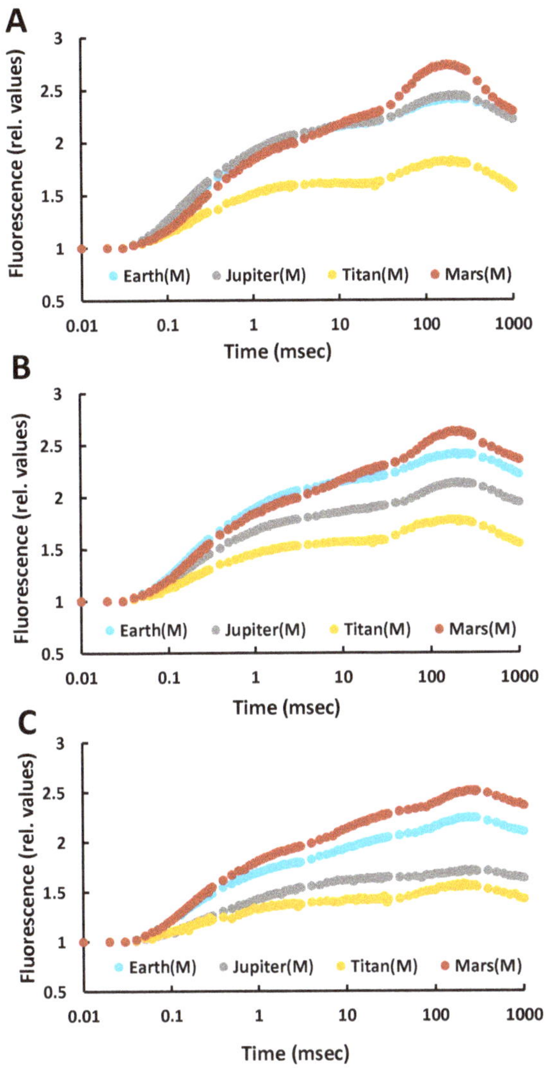

Figure 5. Fluorescence induction curves of chlorophyll for all treatments with mixotrophic cultures of the microalga *Chlorella vulgaris* under different simulated atmospheres (Jupiter, Titan, Mars) compared to the control treatment (Earth atmosphere), after incubation under these conditions for (**A**) 2 days, (**B**) 5 days, and (**C**) 12 days.

In addition to the fluorescence induction curves, we evaluated JIP-test parameters related to the structure and functionality of the photosynthetic mechanism under mixotrophic conditions. This was performed to better determine the functionality of the photosynthetic system in extraterrestrial atmospheres. The values of the parameters were normalized to the control (Earth). The same parameters evaluated under autotrophic conditions were examined, including Fv/Fm, Sm, dV/dt_0, N, TR_0/RC, ET_0/RC, RC/CS_0, ABS/RC, DI_0/RC, and PI(abs). These parameters provided a comprehensive view of the photosynthetic activity of *C. vulgaris* under mixotrophic conditions in different planetary atmospheres.

On Day 2, all parameters were nearly identical to the normalized Earth values (Figure 6A). Several Mars parameters were higher than those of Earth, with the PI(abs) parameter being significantly higher for Mars compared to Earth, while the DI_0/RC ratio was considerably lower. This indicates that the photosynthetic efficiency of the microalga in a Martian atmosphere is high, with low non-photochemical energy dissipation.

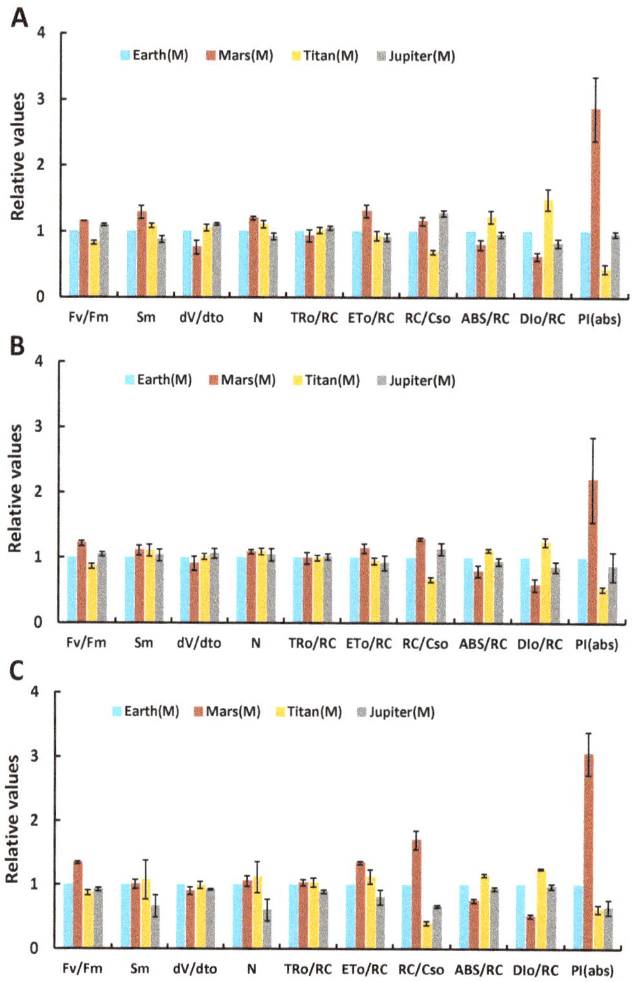

Figure 6. Differentiation of selected parameters from the JIP test (highlighting the structure and function of the photosynthetic mechanism) for treatments under simulated Jupiter, Titan, and Mars atmospheres compared to the control treatment (Earth atmosphere) under mixotrophic conditions, after incubation under these conditions for (**A**) 2 days, (**B**) 5 days, and (**C**) 12 days.

On Day 5, the parameters remained nearly unchanged compared to Earth and Day 2 values. The Sm and dV/dt_0 parameters for the other planets were similar to the control (Earth), indicating a large electron transfer pool and high electron transfer rate. This demonstrates that the photosynthetic mechanism continues to function efficiently (Figure 6B). By Day 12, the deviations among the parameters of the various planets were slightly more pronounced. However, the recorded stress was minimal and significantly less than that observed under autotrophic conditions (Figure 6C).

To confirm the functionality of the microalga's metabolism under mixotrophic conditions, we measured photosynthetic activity (net photosynthesis), expressed as mL O_2·mL PCV^{-1}·day^{-1}, using gas chromatography with a thermal conductivity detector (GC-TCD).

The results showed significant photosynthetic activity in all extraterrestrial atmosphere experiments under mixotrophic conditions (except for Mars, where the extremely low atmospheric pressure made GC-TCD measurements impossible). The highest O_2 production rate was observed in Earth's atmosphere, followed by Titan and then Jupiter (Figure 7). Additionally, photosynthetic rates under all extraterrestrial conditions were higher than those observed under autotrophic conditions (Figure 7).

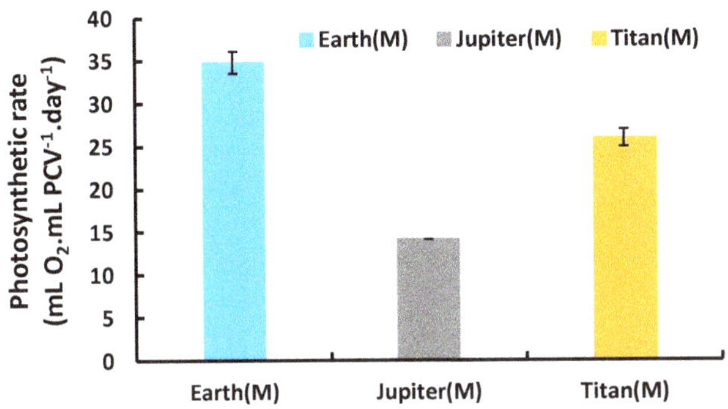

Figure 7. Net photosynthesis rate of *Chlorella vulgaris* under mixotrophic conditions in different atmospheres.

Tracking biomass differentiation over time revealed that *C. vulgaris* cultures grown under mixotrophic conditions in various planetary environments for 2, 5, and 12 days exhibited greater growth than under corresponding autotrophic conditions. This confirms the metabolic functionality of the microalga in all experiments (Figure 8). All measurements on Day 0 began with 1 µL PCV/mL. Over subsequent days, the biomass increased significantly compared to cells grown in autotrophic nutrient media. This is due to the greater energy production enabled by the presence of organic carbon in the medium, leading to higher cell division rates and biomass differentiation between autotrophic and mixotrophic cultures. Biomass growth was observed in all planetary environments, not just Earth, demonstrating that the cells can not only survive but also proliferate in extraterrestrial environments under mixotrophic conditions.

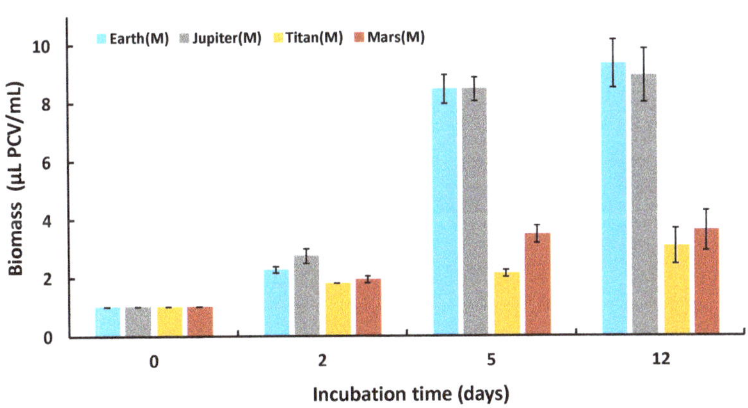

Figure 8. Biomass differentiation kinetics of *Chlorella vulgaris* for all treatments under mixotrophic conditions.

4. Discussion

This study investigated the ability of the unicellular eukaryotic green alga *Chlorella vulgaris* to survive and metabolize by maintaining its photosynthetic processes under various extraterrestrial atmospheric conditions. These included simulated atmospheres resembling those of Mars, Jupiter, and Titan. Using Earth as a control and comparing autotrophic and mixotrophic conditions across different planetary atmospheres, the experimental approach provided valuable insights into the adaptability (without any acclimation period) and metabolic functionality of this microalga in environments vastly different from Earth.

Analysis of the structure and functionality of the photosynthetic machinery of *C. vulgaris* using the JIP-test revealed that the microalga maintained its photosynthetic efficiency under all simulated atmospheres, including those of Mars, Jupiter, and Titan. Fluorescence induction curves indicated that the microalga's photosynthetic efficiency remained largely intact (or only marginally stressed) in all atmospheres. Fluorescence induction curves revealed a substantial decline in fluorescence by day 12 across all autotrophic cultures regardless of atmospheric composition, indicating that while the photosynthetic apparatus remained functional, it was significantly stressed. This decline appears to be a generic effect of the closed culture system rather than a direct result of the atmospheric conditions. In contrast, the mixotrophic cultures exhibited more stable fluorescence patterns, likely due to the supplemental energy and carbon provided by glucose, which mitigated these stresses and sustained growth. The similarity between these curves and those of the Earth-like control suggests that the microalga can survive and operate its photosynthetic machinery, without any acclimation period, in entirely foreign, anoxic planetary environments. Specifically, Fv/Fm values, which represent the maximum photosynthetic efficiency of Photosystem II (PSII), showed no significant photoinhibition or PSII stress in any tested environment.

Particularly intriguing were the results from the simulated Martian atmosphere, where the fluorescence curves and Fv/Fm values surpassed those observed under Earth-like conditions, reflecting the health and high photosynthetic performance of the microalga. These findings are further supported by the higher biomass growth observed in purely photosynthesizing cultures under the Martian atmosphere. The elevated Fv/Fm values and fluorescence curves not only highlight the robust photosynthetic performance of *C. vulgaris*, but also align with its enhanced capacity to efficiently utilize the CO_2-rich, low-pressure environment of Mars for growth. Previous studies have shown that elevated CO_2 concentrations stimulate photosynthesis in microalgae by enhancing carbon fixation rates. Specifically, *C. vulgaris* exhibits peak photosynthetic activity at 30–40% CO_2, while concentrations above 60% inhibit photosynthetic efficiency [2]. However, in conditions approximating 100% CO_2—such as those of the Martian atmosphere—coupled with near-vacuum atmospheric pressure (<10 mbar), the photosynthetic mechanism remained almost unaffected. These findings are consistent with recent studies [10], which suggest that the reduction in atmospheric pressure on Mars may increase thylakoid membrane fluidity, enhancing plastoquinone (PQ) mobility and potentially improving electron transport and photosynthetic activity. However, we also acknowledge that the observed improvement in CO_2 fixation under Martian conditions may be more simply explained by the elevated CO_2 concentrations (even at very low atmospheric pressures), which could directly enhance photosynthetic performance independent of pressure effects.

After extended incubation periods of 5 and 12 days in closed systems, deviations in fluorescence curves became more pronounced, particularly in the atmospheres of Jupiter and Titan, indicating signs of stress. This stress cannot be attributed to a lack of O_2, as these cultures generated their own oxygen through photosynthesis (Figure 3). Instead, it is likely due to imbalances in metabolic processes under these reducing atmospheres.

For instance, CO_2 may be generated via fermentative metabolism, providing a substrate for photosynthesis. In the absence of atmospheric O_2 or CO_2, organic reserves within the microalgae can be metabolized through fermentative pathways, producing CO_2, which is then reinvested in photosynthesis to generate organic matter and O_2. Previous studies [5] suggest that such conditions hinder or reduce photosynthetic efficiency without completely suppressing it. The initial metabolism of these organic reserves, in the absence of atmospheric O_2 and CO_2, appears to rely on nitrogen oxides likely derived from nitrate in the culture medium or other cellular mechanisms, facilitating the production of O_2 through photosynthesis [14,15]. However, a significant portion of the photosynthetically produced O_2 is reinvested in respiration, reducing its release into the atmosphere. This explains the lower photosynthetic activity (expressed as mL O_2·mL PCV^{-1}·day^{-1}) observed under these extraterrestrial atmospheres compared to Earth-like controls, despite the retention of high photosynthetic efficiency (expressed as Fv/Fm). Such metabolic dynamics also contribute to the more limited growth observed under these conditions.

By day 12, fluorescence curves across all autotrophic conditions (including the control) showed increased stress, likely due to the constraints of closed systems. After 12 days of incubation, nutrients began to deplete as they were consumed by the microalgae. Nutrient limitations can reduce electron transport rates, chlorophyll content, and carbon assimilation, explaining the overall decline in photosynthetic efficiency across all atmospheres, including Earth [16]. Despite this stress, *C. vulgaris* continued to survive and photosynthesize, highlighting the resilience of its photosynthetic machinery in extreme environments.

Under mixotrophic conditions, *C. vulgaris* exhibited enhanced photosynthetic activity across all examined planetary atmospheres compared to autotrophic conditions. The presence of an external organic carbon source such as glucose allowed the cells to utilize both photosynthesis and respiration through feedback between these metabolic pathways. This resulted in increased energy production (compared to autotrophic conditions) and, consequently, higher growth rates [17]. Fluorescence curves consistently showed high photosynthetic activity throughout the experiment (up to the 12th day of incubation). Mixotrophic conditions enable *C. vulgaris* to thrive by utilizing both light and organic substrates for energy production, making it particularly suitable for environments where light may be limited or intermittent [18]. Furthermore, the stability of several JIP-test parameters, such as Sm and dV/dt_0, under different planetary atmospheres suggests that the electron transport system remained even more efficient under mixotrophic conditions, resulting in enhanced photosynthetic activity and sustained growth.

Our findings build upon previous investigations into terrestrial life's adaptability to extraterrestrial conditions, which have demonstrated the resilience of methanogenic bacteria [19] and non-methanogenic organisms [20,21] to various atmospheric compositions. Unlike anaerobic systems or those reliant on chemical energy, the photosynthetic activity of *C. vulgaris* highlights the potential of oxygenic photosynthesis to sustain life and recycle atmospheric gases in extraterrestrial habitats. This resilience and adaptability of *C. vulgaris* in extreme (by Earth standards) planetary atmospheres highlights its potential use in bioregenerative life support systems (BLSS). These systems are designed to regenerate oxygen, water, and food while recycling waste to sustain closed-loop ecosystems, making them crucial for long-term human survival in deep space [22,23]. Cultivation of this microalga could contribute to sustained oxygen production in future astronaut colonies on other planets by consuming the CO_2 exhaled by astronauts [2,24]. Additionally, the organism is nutrient-rich and considered a superfood [25]. The rapid biomass production, particularly in mixotrophic cultures where external carbon sources are available, underscores the potential for enhancing microalgal productivity in space habitats where light alone may be insufficient for sustained growth.

This study highlights the resilience of *C. vulgaris* in simulated extraterrestrial atmospheric compositions under laboratory conditions, emphasizing its potential for bioregenerative applications in space exploration. While the initial gas compositions and pressures reflected those of Mars, Jupiter, and Titan, the incubation temperature (25 °C) was maintained to support active photosynthesis and metabolism, differing from the actual environmental temperatures of these celestial bodies. Pressure dynamics within the sealed vials likely shifted during the experiments, altering initial conditions. Unlike studies focusing on survival under extreme low pressures, such as methanogens at 6–143 mbar [26], our research examines active metabolic and photosynthetic responses. These findings complement existing research by expanding our understanding of life's adaptability to extraterrestrial environments.

The findings of this study underscore the versatility of *C. vulgaris*, a common unicellular eukaryotic organism, not only to survive, but also to immediately adapt its photosynthetic mechanism to extraterrestrial environments. The orchestrated ability of this microalga to function in extreme planetary atmospheres, such as those of Mars, Jupiter, and Titan, demonstrates that the microalga can respond with different "metabolic compositions" depending on its environment. These data offer a new perspective on supporting life in space and open new pathways for revisiting the potential transfer of life through the theory of panspermia. This study illustrates that, in the vast universe, even though conditions may appear harsh and the survival of terrestrial microorganisms unlikely, life, as exemplified by *C. vulgaris*, has ways of orchestrating its metabolism to persist in extreme atmospheric conditions [27]. The survival and growth of *C. vulgaris* in non-terrestrial atmospheric conditions redefine the concept of habitable zones and environments, demonstrating that Earth-like life can adapt to a broader range of atmospheric and environmental conditions than previously thought, expanding our criteria for assessing the habitability of exoplanets. However, it is important to note that these results are based on controlled simulations and do not account for the absence of liquid water in such environments.

The discovery that the eukaryotic unicellular green alga *C. vulgaris* can survive and function effectively in the atmospheres of Mars, Jupiter, and Titan holds profound implications for astrobiology, space exploration, and biotechnological applications under extreme conditions.

5. Conclusions

The results of this study demonstrate the remarkable resilience of the microalga *Chlorella vulgaris* in extreme extraterrestrial environments, showcasing its ability to survive, photosynthesize, and proliferate under simulated atmospheric conditions of Mars, Jupiter, and Titan. This resilience, observed under both autotrophic and mixotrophic conditions, underscores the potential of *C. vulgaris* as a key biological component for sustaining life in space. Its capacity to maintain photosynthetic activity and growth in atmospheres vastly different from Earth's highlights its suitability for bioregenerative life support systems, resource utilization, and potential terraforming efforts. Furthermore, the findings contribute valuable insights into the resilience of terrestrial life forms, challenging traditional definitions of habitability and providing a foundation for exploring astrobiological possibilities. By leveraging the unique capabilities of *C. vulgaris*, humanity can take significant strides toward sustainable space exploration, resource-independent missions, and the long-term goal of extraterrestrial colonization.

Supplementary Materials: The following supporting information can be downloaded at: https://www.mdpi.com/article/10.3390/life15010117/s1, Supplementary Figure S1: Preparation of *Chlorella vulgaris* cultures for experimental setup.

Author Contributions: Conceptualization, K.K.; methodology, K.K., A.L. and A.P.; validation, A.L., A.P. and K.K.; investigation, A.L. and A.P.; resources, K.K.; writing—original draft preparation, A.L.; writing—review and editing, K.K., A.L. and A.P.; visualization, A.L.; supervision, K.K.; project administration, K.K. All authors have read and agreed to the published version of the manuscript.

Funding: This research received no external funding.

Institutional Review Board Statement: Not applicable.

Informed Consent Statement: Not applicable.

Data Availability Statement: The original contributions presented in this study are included in the article material. Further inquiries can be directed to the corresponding author.

Acknowledgments: The authors would wish to acknowledge that ChatGPT was used exclusively for linguistic improvements.

Conflicts of Interest: The authors declare no conflicts of interest.

References

1. Buis, A. *The Atmosphere: Getting a Handle on Carbon Dioxide*; NASA: Washington, DC, USA, 2019. Available online: https://science.nasa.gov/earth/climate-change/greenhouse-gases/the-atmosphere-getting-a-handle-on-carbon-dioxide/ (accessed on 9 October 2019).
2. Mountourakis, F.; Papazi, A.; Kotzabasis, K. The microalga *Chlorella vulgaris* as a natural bioenergetic system for effective CO_2 mitigation—New perspectives against global warming. *Symmetry* **2021**, *13*, 997. [CrossRef]
3. Wang, B.; Ye, T.; Li, X.; Bian, P.; Liu, Y.; Wang, G. Survival of desert algae *Chlorella* exposed to Mars-like near space environment. *Life Sci. Space Res.* **2021**, *29*, 22–29. [CrossRef] [PubMed]
4. Wang, B.; Li, X.; Wang, G. Responses of the desert green algae, *Chlorella sp.* to drought stress. *J. Phycol.* **2023**, *59*, 1299–1309. [CrossRef] [PubMed]
5. Zerveas, S.; Kydonakis, E.; Moutidis, P.; Maragkoudakis, A.; Kotzabasis, K. Microalgae strategy in anoxic atmospheres with various CO_2 concentrations–Environmental and (astro)biotechnological perspectives. *Environ. Exp. Bot.* **2021**, *187*, 104474. [CrossRef]
6. Revellame, E.D.; Aguda, R.; Chistoserdov, A.; Fortela, D.L.; Hernandez, R.A.; Zappi, M.E. Microalgae cultivation for space exploration: Assessing the potential for a new generation of waste to human life-support system for long duration space travel and planetary human habitation. *Algal Res.* **2021**, *55*, 102258. [CrossRef]
7. Safi, C.; Zebib, B.; Merah, O.; Pontalier, P.-Y.; Vaca-Garcia, C. Morphology, composition, production, processing and applications of *Chlorella vulgaris*: A review. *Renew. Sustain. Energy Rev.* **2014**, *35*, 265–278. [CrossRef]
8. Ru, I.T.K.; Sung, Y.Y.; Jusoh, M.; Wahid, M.E.A.; Nagappan, T. *Chlorella vulgaris*: A perspective on its potential for combining high biomass with high value bioproducts. *Appl. Phycol.* **2020**, *1*, 2–11. [CrossRef]
9. Nichols, H.W.; Bold, H.C. *Trichosarcina polymorpha* Gen. et Sp. Nov. *J. Phycol.* **1965**, *1*, 34–38. [CrossRef]
10. Gritsi, C.; Sarmas, E.; Daskalakis, V.; Kotzabasis, K. Acclimation mechanism of microalgal photosynthetic apparatus under low atmospheric pressures–new astrobiological perspectives in a Mars-like atmosphere. *Funct. Plant Biol.* **2024**, *51*, FP24058. [CrossRef]
11. Kautsky, H.; Hirsch, A. Neue Versuche zur Kohlensäureassimilation. *Naturwissenschaften* **1931**, *19*, 964. [CrossRef]
12. Bussotti, F.; Desotgiu, R.; Pollastrini, M.; Cascio, C. The JIP test: A tool to screen the capacity of plant adaptation to climate change. *Scand. J. For. Res.* **2010**, *25* (Suppl. S8), 43–50. [CrossRef]
13. Strasser, B.J.; Strasser, R.J. Measuring fast fluorescence transients to address environmental questions: The JIP-Test. In *Photosynthesis from Light to Biosphere*; Mathis, P., Ed.; Kluwer Academic Press: Dordrecht, The Netherlands, 1995; pp. 977–980.
14. Igamberdiev, A.U.; Hill, R.D. Plant mitochondrial function during anaerobiosis. *Ann. Bot.* **2009**, *103*, 259–268. [CrossRef]
15. Gupta, K.J.; Igamberdiev, A.U. The anoxic plant mitochondrion as a nitrite: NO reductase. *Mitochondrion* **2011**, *11*, 537–543. [CrossRef] [PubMed]
16. Liu, T.; Chen, Z.; Xiao, Y.; Yuan, M.; Zhou, C.; Liu, G.; Fang, J.; Yang, B. Biochemical and morphological changes triggered by nitrogen stress in the oleaginous microalga *Chlorella vulgaris*. *Microorganisms* **2022**, *10*, 566. [CrossRef]
17. Cui, H.; Meng, F.; Li, F.; Wang, Y.; Duan, W.; Lin, Y. Two-stage mixotrophic cultivation for enhancing the biomass and lipid productivity of *Chlorella vulgaris*. *AMB Express* **2017**, *7*, 187. [CrossRef]
18. Yun, H.-S.; Kim, Y.-S.; Yoon, H.-S. Effect of different cultivation modes (photoautotrophic, mixotrophic, and heterotrophic) on the growth of *Chlorella sp.* and biocompositions. *Front. Bioeng. Biotechnol.* **2021**, *9*, 774143. [CrossRef]
19. Zeikus, J.G. The Biology of Methanogenic Bacteria. *Bacteriol. Rev.* **1977**, *41*, 514–541. [CrossRef] [PubMed]

20. Seager, S.; Huang, J.; Petkowski, J.J.; Pajusalu, M. Laboratory studies on the viability of life in H2-dominated exoplanet atmospheres. *Nat. Astron.* **2020**, *4*, 802–806. [CrossRef]
21. Berry, B.J.; Jenkins, D.G.; Schuerger, A.C. Effects of Simulated Mars Conditions on the Survival and Growth of *Escherichia coli* and *Serratia liquefaciens*. *Appl. Environ. Microbiol.* **2010**, *76*, 2377–2386. [CrossRef]
22. Cycil, L.M.; Hausrath, E.M.; Ming, D.W.; Adcock, C.T.; Raymond, J.; Remias, D.; Ruemmele, W.P. Investigating the growth of algae under low atmospheric pressures for potential food and oxygen production on Mars. *Front. Microbiol.* **2021**, *12*, 733244. [CrossRef]
23. Liu, H.; Yao, Z.; Fu, Y.; Feng, J. Review of research into bioregenerative life support system(s) which can support humans living in space. *Life Sci. Space Res.* **2021**, *31*, 113–120. [CrossRef] [PubMed]
24. Detrell, G. *Chlorella vulgaris* photobioreactor for oxygen and food production on a moon base—Potential and challenges. *Front. Astron. Space Sci.* **2021**, *8*, 700579. [CrossRef]
25. Yang, L.; Li, H.; Liu, T.; Zhong, Y.; Ji, C.; Lu, Q.; Fan, L.; Li, J.; Leng, L.; Li, K.; et al. Microalgae biotechnology as an attempt for bioregenerative life support systems: Problems and prospects. *J. Chem. Technol. Biotechnol.* **2019**, *94*, 3039–3048. [CrossRef]
26. Mickol, R.L.; Kral, T.A. Low Pressure Tolerance by Methanogens in an Aqueous Environment: Implications for Subsurface Life on Mars. *Orig. Life Evol. Biosph.* **2017**, *47*, 511–532. [CrossRef] [PubMed]
27. Angelakis, G.N.; Psarologaki, C.; Pirintsos, S.; Kotzabasis, K. Extremophiles and Extremophilic Behaviour—New Insights and Perspectives. *Life* **2024**, *14*, 1425. [CrossRef] [PubMed]

life

Article

Circular Utilization of Coffee Grounds as a Bio-Nutrient through Microbial Transformation for Leafy Vegetables

Hasan Ozer [1], Naime Ozdemir [2], Asude Ates [1,*], Rabia Koklu [1], Sinem Ozturk Erdem [2] and Saim Ozdemir [1]

[1] Department of Environmental Engineering, Faculty of Engineering, Sakarya University, Esentepe, 54187 Sakarya, Turkey; hasanozer@sakarya.edu.tr (H.O.); rkoklu@sakarya.edu.tr (R.K.); saimo@sakarya.edu.tr (S.O.)

[2] Department of Horticulture, Faculty of Agriculture and Natural Science, Bilecik Seyh Edebali University, 11230 Bilecik, Turkey; naime.ozdemir@bilecik.edu.tr (N.O.); sinem.erdem@bilecik.edu.tr (S.O.E.)

* Correspondence: aates@sakarya.edu.tr

Abstract: This study explores the production of bio-nutrients from bioactive compound-rich spent coffee grounds (SCG) and biochar (BC) through composting after inoculation with a biological agent and its impact on the growth performance of garden cress and spinach. The SCG was composted with six doses of BC (0, 5, 10, 15, 20, and 25%). The compost with 10% BC exhibited the best maturity, humification, and phytotoxicity index values of dissolved organic carbon (DOC), humification index (E4/E6), and germination index (GI). A metagenome analysis showed that compost starter enhanced the bacterial community's relative abundance, richness, and diversity in SCG and BC treatments. This improvement included increased *Patescibacteria*, which can break down noxious phenolic compounds found in SCG and BC. The BC enriched the compost with phosphorus and potassium while preserving the nitrogen. In plant growth experiments, the total chlorophyll content in compost-treated garden cress and spinach was 2.47 and 4.88 mg g^{-1}, respectively, which was significantly greater ($p \leq 0.05$) than in unfertilized plants and similar to the plants treated with traditional fertilizer. Overall, the results show that the compost of SCG + BC was well-suited for promoting the growth of garden cress and spinach, providing adequate nutrients as a fertilizer for these leafy vegetables.

Keywords: poultry litter biochar; starter inoculation; plant nutrients; spinach; garden cress

1. Introduction

In today's world, there is a significant amount of disposable bio-waste that must be effectively managed to align with the United Nations Sustainable Development Goals. Failing to do so could lead to considerable environmental damage and loss of valuable nutrients [1]. Circular economy models offer an array of valorization methods to mitigate the adverse effects of bio-waste by keeping resources in the economy for as long as possible and minimizing waste. These approaches enable the reutilization of resources, thus creating additional value [2]. One widely used biotechnological method is composting, which involves converting various organic raw materials, including harmful compounds found in spent coffee grounds (SCG), into nutrient-rich biofertilizers through a diverse and beneficial microbial community. This biofertilizer can then be reused in crop production, promoting carbon neutrality [3] and supporting the transition to recyclable technology and domestic resource security [4].

Spent coffee grounds (SCGs) and the thermochemical process by-product biochar are the latest of decades of research into the full or partial replacement of peat with other organic and inorganic materials [5]. Both SCG and biochar are sterile to microbes due to high temperature production and contain stable carbon and plant nutrients, making them an interesting material for research topics from an agri-environmental perspective [6]. Similar to other bio-wastes, SCG is mainly disposed of in landfills and thus poses a threat to the environment as it contains useful plant nutrients and chemical compounds of biotechnological interest, such as organic acids, tannins, and phenolic compounds [7]. Although SCG contains significant amounts of plant nutrients, the direct use of SCG for soil improvement is associated with limitations due to its fine nature, low C/N ratio, and high moisture content, such as clogging of soil porosity, hydrophobicity of soil aggregates, phytotoxicity, and nutrient imbalance for crops [8] even for plants lower in acid. In this context, bioconversion through composting was proposed as a safe practice to improve challenges of SCG in order to fully exploit its nutritional value and soil mitigating potential.

Combining acidic spent coffee grounds (SCGs) with alkaline biochar through composting could present a sustainable solution for utilizing high-nutrient waste in the preparation of bio-nutrient or growing media constituents. Composting is a safe and straightforward process that contributes to the circular economy model and efficient waste management [9], enabling renewability in crop production. However, raw SCG and biochar do not possess the necessary characteristics for ideal growing media due to their structural, biological, and chemical limitations [10]. For example, SCG contains various bioactive compounds, such as polyphenols, tannic acid, organic acids, lipids, and caffeine, which can be phytotoxic to certain plants [11]. Similarly, although biochar offers benefits such as carbon sequestration, acting as a nutrient source and retaining nutrients in the soil–plant matrix, its alkaline nature restricts its direct use, potentially causing toxic effects or nutrient imbalances, thus requiring conditioning before use [12]. By leveraging the potentially favorable chemical composition of SCG and the porous physical properties of biochar, favorable composting conditions can be created, providing a natural organic nutrient source for potted crops [13]. Composting is a microbial-driven process that rapidly breaks down readily available organic matter into humus, yielding several organic acids that are essential for optimizing both materials for crop fertilization.

In recent years, biochar was found to be capable of enhancing biological processes, increasing bacterial diversity by providing a promising microbial environment for several forms of aerobic and anaerobic degradation, particularly for composting microorganism [14]. In this regard, the use of biochar as a microbial-enhancing tool would be a highly desired for rapid composting processes, where noxious bioactive compounds present in SCG and BC can be converted into nonphytotoxic plant nutrients that are safely applicable to cultivated crops.

Setting up new circular economic approaches in waste management in urban ecosystems by valorizing by-products of the agro-food industry as a secondary feedstuff for the crop production cycle is crucial for the sustainability of the urban farming and the environment. The present research aimed to examine the feasibility of producing a nutrient-rich bio-nutrient source by composting spent coffee grounds with poultry litter biochar and evaluate their potential use as a nutrient resource in pot growing media for leafy vegetable production in an urban vertical growing system setting. It was hypothesized that composting SCG with biochar would be a suitable growing media option for nutrient-rich horticultural substrates. To test this hypothesis, pot experiments were performed in a vertical growing system using two leafy vegetable crop garden cresses and spinach employing different percentages of compost to replace peat-based growing substrates.

2. Materials and Methods

2.1. Composting and Growing Media Preperation

The raw materials used in the compost experiments consisted of spent coffee grounds collected from the coffee shop market Sakarya, Turkey. Biochar was provided by a commercial gasification plant (Bolu province, Turkey) that uses poultry litter collected from a nearby poultry farm with wood shavings/saw dust bedding material for chicks. The gasification plant process wastes at 600 °C with a residence time of 3.5 h to produce mainly syngas for electricity generation. At the same time, several byproducts are generated, such as biochar. The main characteristics of SCG and BC based on elemental analyzer (LECO CHN628, St. Joseph, MI, USA) in accordance with ASTM D5373-14 guidelines are presented in Table 1. The pH, total nitrogen content (TN), and total organic carbon (C) of SCG were 5.61, 2.16, and 52.06%, respectively. The main physical characteristics of poultry litter biochar were as follows: alkaline in nature, particles of 2–4 mm, the Brunauer–Emmett–Teller (BET) surface area of 13.58 m^2 g^{-1} and total pore volume of 1.55 cm^3 g^{-1}, and average pore diameter 6.21 Å) (Figure 1). It is rich in non-volatile essential nutrients for plant fertilizer such as P and K (Table 1, Figure 1).

Table 1. Properties of raw spent coffee ground and poultry litter biochar as received form.

Parameters	Spent Coffee Ground	Biochar
Moisture content (%)	63.62	10.26
Ash content (%)	3.15	35.82
Volatile matter (%)	96.85	51.43
pH	5.61	10.25
C %	52.06	52.25
H %	5.87	2.04
N %	2.16	1.27
S %	0.20	0.05
O %	58.17	44.39
C/N	24.08	23.12

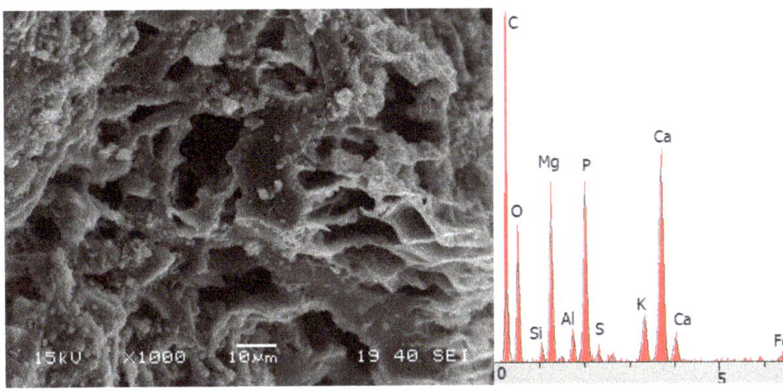

Element	%
C	28.466
O	26.978
Na	0.851
Mg	7.554
Al	0.480
Si	2.594
P	7.251
S	1.359
K	5.471
Ca	15.761
Fe	3.235
Total	100.000

Figure 1. SEM image and EDS analysis of poultry litter biochar.

The experiment used small 6 L insulated reactors (with a diameter of 27.5 cm and a height of 17 cm) for 45 days. Different compost treatments were mixed with BC at varying percentages (5%, 10%, 15%, 20%, and 25% dry basis v/v) to create biochar-amended SCG composts. The control treatment used SCG without BC amendment. All the treatments were mixed with bacterial aliquots, and the moisture content was adjusted to 60%. The mixture (about 3 kg) was then placed into compost reactors with a hole at the bottom for leachate drainage or airflow.

Microorganisms from well-matured maize straw compost were extracted to serve as a starter for SCG, BC compost samples, which are initially sterile due to exposure to high temperatures. The mature compost was sourced from farm-based windrow composting facilities of maize straw that were curing for eight months, making it an excellent source of microorganisms. The process involved mixing 100 g of compost samples with 1 L of distilled water and shaking the mixture for 2 h to produce eluates rich in composting bacteria. The resulting solution was then evenly sprayed onto the compost materials as a composting starter.

The compost mixtures were mechanically aerated daily for the first two weeks and every third day for the following 6 weeks to ensure oxygenation and homogenization. The mixing process was carried out rapidly to reduce heat and evaporation loss. Temperatures initially reached the thermophilic range (>55 °C) and remained there for the first week.

After that, temperatures gradually decreased to ambient temperatures (20–25 °C) in the following weeks. After 4 weeks of composting, the mixtures were left open during the curing stage to reach maturity. Compost samples were collected after four weeks to determine maturity and chemical characteristics.

2.2. Microbial Community Analysis in Compost Samples

The composted samples collected at the end of the composting experiments from the, inoculated SCG compost and inoculated SCG + BC10 were used for microbial community analysis compared to the uninoculated SCG compost. The 16S rRNA gene was amplified and analyzed using QIIME 2. Following amplification with specific primers, purification was carried out to create the library. Illumina binary indexes and adapters were added in the index PCR step using the Nextera XT index kit, followed by another round of purification. The library concentration was measured using real-time PCR (Kyratec, SC-300G, SuperCycler, Caboolture, QLD, Australia) and then diluted to 4 nM before normalization. Normalized samples were pooled together. During sequencing, fluorescence of the added base was optically observed and recorded each time a new deoxyribose nucleotide triphosphate (dNTP) was incorporated using the synthesis method. The raw data generated after sequencing were converted to FASTA format for analysis. It is important to note that the FASTA format is used to encode nucleotide or amino acid sequences, and raw sequence data in FASTA or FASTQ format can be directly used in next-generation microbiome bioinformatics platforms such as QIIME. The detailed analytical procedures were described by [9]. The sequence analysis of the compost obtained from the PCR products and the reporting of the microbial communities were performed by the laboratory of BM Labosis.

2.3. Physical and Chemical Analysis

For a phytotoxicity test, seed germination and germination index (GI%) were estimated using garden cress (*Lepidium sativum* L.) obtained from local market. Briefly, a solution of 2.5 g dw of compost in 25 mL of water was prepared, shaken for 2 h, then centrifuged at 3000 rpm, and filtered by Whatman 2 filtering paper. Ten seeds were placed on sterilized filter paper in a Petri dish, with three replicate dishes for each growing compost. Afterwards, 5 mL of the compost solution was added to the corresponding Petri dishes. Similarly, a control was prepared with deionized water. All the Petri dishes were incubated in the dark for 72 h at 25 °C. GI was calculated from the relative seed germination % and relative root length % ($GI (\%) = 100 \times G/Gc \times L/Lc$) [15], where G and L stand for the germinated seeds and root length of the samples, respectively, and Gc and Lc stand for the corresponding values for the distilled water. For relative germination, living seedlings that germinated were counted. Similarly, for relative root length, living roots were measured.

In order to measure dissolved organic carbon (DOC), the samples were treated with ultra-pure water at a solid to water ratio of 1:10 (*dw/v*), using a horizontal shaker at 200 rpm for 3 h, followed by centrifugation at $3000\times g$ for 20 min and filtration with 0.45 µm syringe filters. The DOC of the leaching solution was analyzed using a TOC analyzer (Shimadzu, VCSH, Kyoto, Japan).

The pH values in the compost samples were determined in water extracts obtained with a solid to liquid ratio of 1:10 (*dw/v*) using a pH meter (Schott CG 840, Eindhoven, The Netherlands).

To determine the degree of humification, a 1 g air-dried sample was agitated in a 50 mL, 0.5 M sodium hydroxide (NaOH) solution for 2 h, followed by centrifugation at 3000 rpm for 25 min. The absorbance of the supernatant was measured using a spectrophotometer (Hewlett–Packard Company, Wilmington, DE, USA) over a range of 190 to 1100 nm to assess the decomposition level. The optical density of the solution was recorded at 465 nm (E4) and 665 nm (E6), and the E4/E6 ratio was then determined to characterize the intensity of decomposition.

The essential plant nutrients and trace element quantifications of the composts, substrates, were conducted using an Inductively Coupled Plasma Optical Emission Spectrometry (ICP-OES) (Spectro Arcos, Kleve, Germany) after nitric (HNO_3) and perchloric acid ($HClO_4$) digestion at 200 °C [15].

For leaf chlorophyll content estimation, plant leaves that are picked directly from the plant at the end of the 30 days are used. The obtained chlorophyll content in plant leaves under the respective treatment is assessed according to the methodology of Ozdemir et al. (2021) [16], using 100 mg of fresh leaf extracted overnight with 80% acetone and centrifuged at 10,000× *g* for 5 min. Then, the solution was put into a cuvette and its absorbance was measured at wavelengths of 645, 663, and 470 nm using a UV–visible spectrophotometer (Hitachi-U2001, Tokyo, Japan).

2.4. Plant Cultivation Experiments

To evaluate the potential fertilizing capacity of the SCG composts, four growing media were prepared by mixing SCG-BC10 compost as fallows: commercial peat without compost (T_0); commercial peat + 15% SCG-BC (T_1); commercial peat + 30% SCG-BC (T_2); commercial peat + 50% SCG-BC (T_3); and commercial peat with fertilizer (T_4). Garden cress (*L. sativum*) and spinach (*Spinacia oleracea* L.) were chosen for the plant production experiment because they germinate quickly, grow fast, are easy to measure, and respond to organic nutrients. The seeds were planted directly into 7 × 7 cm pots and the pots were placed on the vertical trellis growth system. The pots were manually watered daily by spraying tap water to each pot until reaching field water capacity to prevent drainage. Emergence was observed after 2 days for garden cress and after 5 days for spinach. After 30 days of emergence, the plant growth experiments were terminated and fresh leaves were collected at the end of the experiments to determine the chlorophyll content.

2.5. Statistical Analysis

An analysis of variance (ANOVA) was performed to assess the variations among the physical and chemical properties of the composts and their agronomic potential. When the *p*-value for the ANOVA was less than 0.05, the means were compared using Tukey's test at a 0.05 significance level. All statistical analyses were carried out using the StatsDirect software package (V2.7.2, StatsDirect, Ltd., Altrincham, Cheshire, UK).

3. Results and Discussion

3.1. Chances in Compost Maturity Parameters

The dissolved organic carbon (DOC) is a key component of the organic matter pool that induces rapid decomposition, and its decrease indicates good microbial activity, effective composting, and high-quality compost [17]. Figure 2a demonstrates the impact of adding biochar on the DOC levels. The DOC content reduced from 5.63 mg kg^{-1} in SCG to 1.11 mg kg^{-1} in the treatment with 25% biochar. Furthermore, the DOC levels decreased during composting, particularly at higher biochar rates. The recommended DOC content for mature compost is 4–10 mg kg^{-1} [18]. According to these values, the addition of biochar accelerated the composting process, with biochar-amended compost reaching stability by week 4, while the SCG control compost stabilized after week 6. The porous nature of biochar decreased compost density, improved aerobic conditions, offered a habitat for microorganisms, and enhanced microbial activity, leading to DOC decomposition [17]. Moreover, the high porosity and large surface area of biochar allowed for the absorption of DOC, contributing to the decrease in its concentration. Importantly, the DOC levels observed in this study met the stability and maturity requirement (<10 g kg^{-1}) as reported in previous research [18].

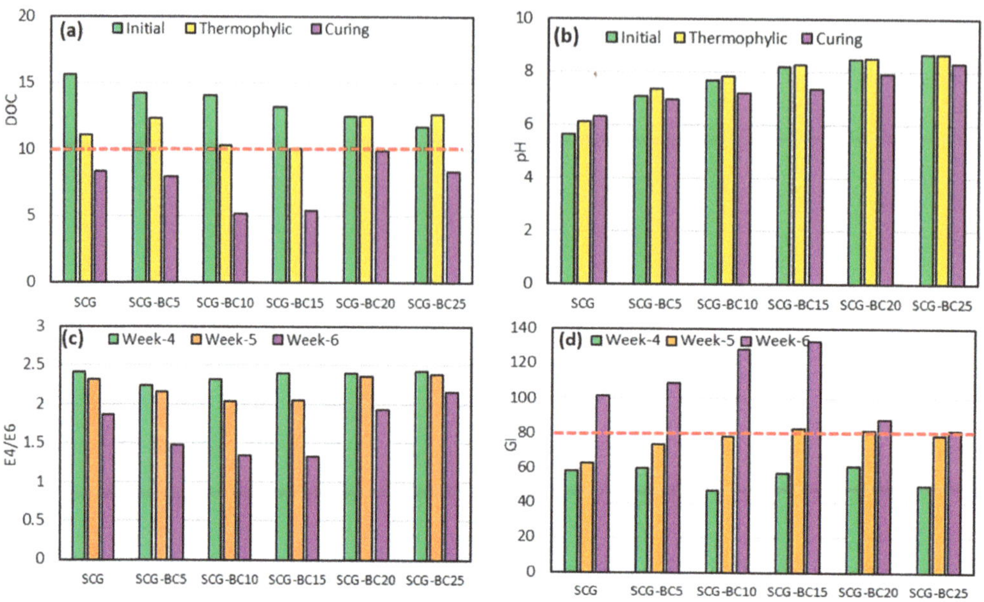

Figure 2. Variations in (**a**) DOC, (**b**) pH, (**c**) E4/E6, and (**d**) GI during composting of spent coffee grounds with various levels of biochar. The red line indicates DOC's stability boundary and GI's phytotoxicity boundary.

PH is a critical factor influencing microbial growth and is commonly used to assess compost maturity. The addition of biochar had a significant effect on pH. The pH values of SCG-BC increased from 6.34 to 8.35 when 5% or 25% biochar was added to the SCG. Despite the acidic nature of fresh SCG, the addition of biochar increased the pH of the compost, which could be related to the formation of alkali metals such as P, K, and Ca, which are abundant in poultry litter [9]. In Figure 2b, it can be seen that the pH values of SCG and SCG-BC decreased significantly during the initial phase, increased during the thermophilic phase, and finally stabilized during the curing phase. The pH of the compost was lowered during the initial rapid decomposition phase by the formation of organic acids and then stabilized by the removal of ammonia [6], the development of humic substances [19], or buffering by biochar [5] in all composts. Maintaining an appropriate pH is essential to promote effective microbial activity and the smooth running of the composting process. The standard for organic fertilizers and compost in Turkey specifies that the pH of the compost should be in the range of 5.5 to 8.5. All compost treatments reached the specified pH range after 6 weeks, indicating the maturity of the compost samples.

Spectroscopic analysis allows us to assess how mature compost is by examining the transformation properties of the organic matter. The E4/E6 ratio, measured by the optical density of compost extract solution at 465 and 665 nm, reflects the degree of humification of the compost. A low E4/E6 value means that the compost is highly mature and stabilized [18]. In the study, it was found that the E4/E6 values decreased significantly when BC was added to the compost at 10 and 15%, indicating that adding BC can improve compost maturity (Figure 2c). However, the decrease in E4/E6 values was slower at 20 and 25% BC-added compost, possibly due to the relatively high pH over 8.5. The quality of the compost was improved due to the stimulation of microbial abundance and biomass, as well as the provision of habitat by adding BC, leading to increased condensation of aromatic humus components during composting.

Germination index (GI) is a sensitive parameter used to evaluate the maturity and phytotoxicity of compost. A GI value of over 80% is necessary to meet maturity standards.

At the end of the fourth week of composting, GI values ranged from 47 to 61%, indicating the presence of phytotoxic substances and immaturity of the compost (Figure 2d). By the fifth week, the GI in compost samples improved and reached the maturity threshold of 80%. After the sixth week, the GI values of the final compost ranged between 81.46% and 132.46% (Figure 2d), indicating complete maturity. The increasing GI values correlated with the rate of biochar addition, showing a peak at 15% biochar, with the highest GI value of 132.46%, a 30.48% increase compared to the SCG. This increase signifies reduced phytotoxicity, attributed to the adsorption and reduced production of phytotoxic substances due to biochar addition. This shows the efficacy of the biochar acting as a bulking agent, also being able to adsorb volatiles and toxic substances. However, at higher biochar levels of 20% and 25%, the GI value decreased (Figure 2d), likely due to the presence of harmful compounds in biochar, such as polycyclic aromatic hydrocarbons, polychlorinated biphenyls, and other toxic compounds [20].

3.2. Bacterial Community and Diversity

Bacteria are the predominant group in composting processes as they have the ability to attack highly complex, versatile organic substrates by releasing a broad spectrum of extracellular enzymes. Therefore, their characterization based on their diversity and richness is fundamental for the understanding of the composting process and especially for the composting of bioactive molecules such as SCG. The roasting, brewing, and cooking of SCG enriches a variety of bioactive compounds in SCG [21]. In the present study, uninoculated SCG, inoculated SCG, and inoculated SCG + BC10 compost samples were subjected to bacterial community and diversity index analysis after six weeks in the maturation phase.

Figure 3 illustrates the impact of starter inoculation on the bacterial community composition at the phylum level and the alpha diversity indices in spent coffee grounds and spent coffee grounds biochar at a 10% mixture. At the end of composting, the compost treatments showed different relative abundance of bacterial communities at the phylum level. There were five dominant phyla in the studied treatments: *Bacteroidota* (25–37%), *Proteobacteria* (19–25%), *Planctomycetota* (5–17%), *Firmicutes* (3–16%), and *Actinobacteriota* (2–11%) (Figure 3a), which is common for compost experiments [22]). However, the five dominant strains were detected in all treatments. Previous studies showed that the sequence of the bacterial community during the composting process is as follows: *Proteobacteria* in the initial phase, *Firmicutes* in the thermophilic and mesophilic phases, and *Bacteroidota* in the maturation phase [23]. On the other hand, inoculation with compost starter increased the relative abundance of *Firmicutes* and *Actinobacteriota* in both the SCG and SCG-BC composting processes, which replaced the *Bacteroidota* and *Planctomycetota* compared to the non-inoculated treatment. These dominant bacterial phyla collectively have a strong ability to decompose organic waste and are usually the most abundant phylum in organic-rich environments [9]. However, the presence of *Firmicutes* and *Actinobacteriota* in higher abundance in the inoculated treatments could apparently be due to a more mature compost. *Firmicutes* and *Actinobacteriota* are prevalent in mature corn straw compost [24], suggesting that the success of starter inoculation and the degradation and maturity it promotes in the final compost. The presence of *Patescibacteria*, which are capable of hydrolyzing phenolic compounds [25], in the inoculated treatments compared to the non-inoculated compost, indicates the effectiveness of the compost starter. In addition, the presence of *Actinobacteria* in the maturation phase indicates a high conversion of lignocellulose to humus [14].

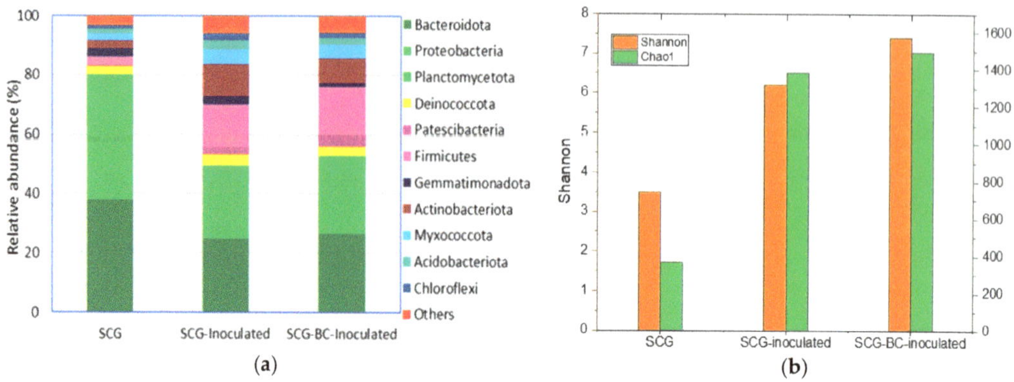

Figure 3. Effect of starter inoculation on compost bacterial community composition at the phylum level (**a**) and, alpha diversity indices (**b**) in spent coffee grounds and spent coffee grounds biochar at a 10% mixture.

The SCG is a rich source of bioactive compounds, such as phenolics (i.e., chlorogenic, gallic, caffeic, quinic, ellagic, cinnamic, coumaric, and ferulic acids), alkaloids (i.e., caffeine, trigonelline), and flavonoids (i.e., catechin, epicatechin, rutin, quercetin, and naringin) [26]. These compounds possess potent biological properties and can function as antimicrobial, antifungal, and antiviral agents [11]. Thus, it is crucial to promote the proliferation of active microbial communities during composting, as these communities play a vital role in neutralizing harmful substances present in SCG. Alpha diversity is utilized to assess the richness and relative abundance of bacteria in compost samples (Figure 3b). The bacterial community diversity and richness of mature compost were notably higher in SCG and SCG + BC. This increase in diversity and richness of the bacterial communities was attributed to bacterial inoculation [27]. The Chao1 index serves as an estimator of species richness, reflecting the total number of distinct species in each sample. Microbial inoculation significantly elevated the Chao1 values in SCG and SCG-BC compost, indicating a greater species richness. SCG-BC exhibited the highest Chao1 value (1496), followed closely by SCG (1454), suggesting that bacterial inoculation significantly boosted microbial species richness. Furthermore, the Shannon index evaluates species diversity and evenness within each sample, with higher values indicating a more balanced distribution of species abundance [28]. SCG-BC and SCG displayed relatively higher Shannon index values (7.40 and 6.20, respectively) compared to uninoculated SCG (3.49), suggesting a more balanced representation of microbial species in inoculated composts [29]. The Shannon diversity index integrates species richness and evenness, with higher Shannon values indicating increased diversity and evenly distributed microbial communities in compost samples. Overall, the alpha diversity results indicate that bacterial inoculation derived from mature compost significantly impacts the creation of diverse microbial communities in SCG and BC that initially did not contain microorganisms. According to the Shannon and Chao1 index, both SCG and SCG-BC exhibited higher diversity than uninoculated SCG, indicating a more varied microbial composition. Therefore, it is possible to expect more rapid composting, as well as the removal of bioactive compounds in spent coffee grounds that have a negative impact on compost maturity and phytotoxicity parameters.

3.3. Effects of Biochar on Nutrient Contents of Composting Products

The variation in essential plant nutrients in the final compost product affected by biochar addition are shown in Table 2. In this study, the content of TN of the final BC-compost products were 2.21–2.88%, which is significantly smaller than the TN content of the SCG compost without biochar amendment (3.20%). Some previous studies showed that the TN contents of the final compost depends on type and rate of bulking agent mixed to compost. Some studies also report that biochar addition had a significant effect on the TN content due to preserving TN in compost [15]. The response of final compost TN to biochar addition was significantly influenced by composting conditions, such as compost method, initial carbon to nitrogen ration, and initial moisture content [30], which may vary dramatically among different material and studies.

Table 2. Nutrient content of increased dose of biochar added spent coffee ground composts (mean ± sd).

Parameter	SCG	SCG-5%BC	SCG-10%BC	SCG-15%BC3	SCG-20%BC	SCG-25%BC
TN (%)	3.20 ± 0.43 a	2.88 ± 0.22 b	2.72 ± 0.11 c	2.61 ± 0.22 d	2.46 ± 0.04 e	2.21 ± 0.02 f
TP (%)	0.20 ± 0.02 f	0.28 ± 0.02 e	0.39 ± 0.01 d	0.42 ± 0.02 c	0.47 ± 0.04 b	0.53 ± 0.01 a
TK (%)	0.49 ± 0.03 f	0.54 ± 0.01 e	0.66 ± 0.06 d	0.71 ± 0.02 c	0.79 ± 0.01 b	0.84 ± 0.02 a
Ca (%)	0.14 ± 0.05 f	0.18 ± 0.03 e	0.24 ± 0.04 d	0.32 ± 0.02 c	0.37 ± 0.03 b	0.43 ± 0.03 a
Mg (%)	0.10 ± 0.02 e	0.11 ± 0.02 d	0.13 ± 0.01 d	0.16 ± 0.02 c	0.19 ± 0.01 b	0.21 ± 0.01 a
S (%)	0.01 ± 0.01 d	0.02 ± 0.01 c	0.02 ± 0.01 c	0.03 ± 0.01 b	0.03 ± 0.05 b	0.04 ± 0.04 a
Fe (mg kg^{-1})	124 ± 1.21 f	136 ± 1.21 e	144 ± 1.21 d	149 ± 1.21 c	153 ± 1.71 b	157 ± 1.23 a
Mn (mg kg^{-1})	58 ± 0.04 f	64 ± 0.89 e	67 ± 1.01 d	72 ± 0.82 c	76 ± 1.03 b	78 ± 0.92 a
Zn (mg kg^{-1})	85 ± 0.91 f	88 ± 0.79 e	92 ± 0.85 d	95 ± 0.87 c	97 ± 0.92 b	103 ± 0.84 a
Cu (mg kg^{-1})	26 ± 0.68 f	40 ± 1.15 e	43 ± 0.93 d	46 ± 1.33 c	50 ± 0.86 b	52 ± 1.44 a
B (mg kg^{-1})	14 ± 0.01 c	15 ± 0.24 c	15 ± 0.34 c	17 ± 0.37 b	17 ± 0.22 b	19 ± 0.51 a

For each row, different lowercase letters indicate significant differences among treatments according to the Tukey test ($p < 0.05$).

In the presence of biochar, the total phosphorus and potassium contents in compost increased from 0.20 to 0.53%, 0.49 to 0.84%, respectively, and significantly higher than with sole SCG control compost (Table 2). This increase could be attributed to the higher levels of P and K in poultry litter BC (Table 1). Previous studies also indicated that compost mixed with biochar can effectively boost essential plant nutrients [31,32]. Compost users criticized the low nutrient content of composts, stating that a large amount of compost is required to meet plant requirements [9]. In this context, nutrient-rich poultry litter BC as a compost additive can satisfy user expectation. Moreover, there is evidence suggesting that organic nutrient sources, when applied at maximum yields, can lead to organic build-up, acting as a slow-release fertilizer due to beneficial microbial action, thus promoting plant growth [15].

The impact of adding biochar on the levels of major plant nutrients N, P, and K is crucial for the quality and potential applications of compost products to cultivated crops. When 10% biochar was added to the final composting product, the P content almost doubled compared to the sole SCG compost, and the K content similarly increased almost twofold. However, the addition of biochar led to a reduction in TN content at each increased level of added BC, ranging from 2.88 to 2.21% (Table 2). This effect is likely due to biochar addition improving the P and K in compost mixture while mitigating N loss through volatilization, attributed to its favorable structure and aeration of the compost substrate, driven by characteristics such as porosity, large specific surface area, and functional nutrient binding groups [13]. Furthermore, alongside the enhancement of plant nutrients, the desired promotion of nutrient mineralization dynamics by microbial enzymes in the compost matrix is typically sought to enhance compost quality and nutrient availability [9]. Therefore, these results confirm the hypothesis of this study that composted SCG can be used as a low-cost and environmentally friendly bio-nutrient resource due to its high nitrogen content and sufficient essential macro and micro plant nutrients.

3.4. *Leafy Vegetable Pigment Contents*

The chlorophyll content in leaves serves as a significant indicator of photosynthetic activity, which in turn contributes to enhanced plant growth as a result of the positive response to vital plant nutrients provided by applied composts [16]. The levels of chl-a, chl-b, and total chlorophyll in garden cress leaves increased with the application of increasing amounts of SCG-BC compost in the growing media (Figure 3a). Notably, plants fertilized with inorganic fertilizer exhibited the highest chl-a and total chlorophyll content, with no significant difference observed compared to compost-treated plants at 30 and 50% compost mixtures. Conversely, unfertilized plants showed the lowest chlorophyll content. The increased leaf chlorophyll content, compared to the control group, may be attributed to the improved nutrient supply provided by the nutrient pool of SCG-BC [33], thereby indirectly indicating enhanced photosynthetic activity. Garden cress, being a leafy vegetable, is particularly receptive to nutrients such as nitrogen, and organic amendments are effective in providing nutrients due to their slow-release properties, as opposed to inorganic nutrients [12].

The levels of chlorophyll a and b, as well as the total chlorophyll (Chl a + b) content in spinach leaves per unit of fresh weight, were significantly influenced by the treatments ($p < 0.01$), indicating a positive response of spinach to the increasing dose of SCG + BC compost. Specifically, the Chl-a content in the leaves was significantly higher in the fertilized treatment (Figure 4b). Among the treatments, T_3 (50% compost) resulted in the highest chlorophyll content in spinach leaves compared to the other compost treatments, which received 15 and 30% compost. These findings demonstrate that the nutrient content and mineralization rate support spinach growth without the need for additional mineral fertilizer, which is crucial for sustainable leafy vegetable cultivation in the short term and reducing the environmental impact of excessive use of inorganic fertilizers. Numerous previous studies demonstrated the sufficiency of organic nutrient sources in potted cultivation due to the improved nutrient use efficiency [34], although some studies indicated the necessity of additional inorganic fertilizers to support plant growth and achieve higher yields [35]. This may be attributed to the limited mineralization rate of organic nutrients from bio-nutrient resources [36]. However, in this study, the nutrient mineralization plant-providing rate of SCG and the nutrient retention capacity of BC proved to be effective in meeting the plant's nutrient demands during a one-month growth cycle. Nonetheless, longer crop growth cycles may require additional fertilizer, as suggested in previous studies that recommend combining compost with fertilizer to maintain optimum tissue nutrient levels, leaf chlorophyll content, and ultimately achieve higher yields [9,18].

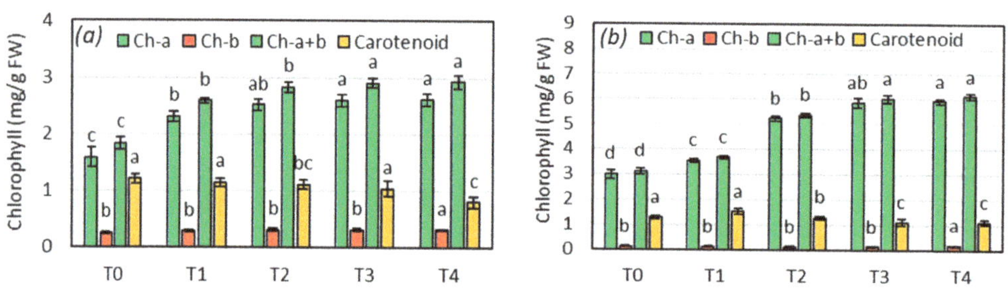

Figure 4. Chlorophyll and carotenoid contents of *L. sativum* (**a**) and *S. oleracea* (**b**) grown in increasing amounts of SCG + BC compost. Values are means for triplicate samples. Different lowercase letters on the histogram of the same color indicate significant differences at $p < 0.05$.

4. Conclusions

The outcomes of composting experiments indicate that it is possible to produce nutrient-rich compost suitable for crop fertilization by combining spent coffee grounds and poultry litter biochar at a 10% ratio and introducing the mixture to starter inoculum obtained from maize straw compost. The incorporation of biochar notably increased the abundance, richness, and diversity of microorganisms involved in composting, including *Patescibacteria*, which are crucial in breaking down phenolic compounds found in spent coffee grounds and biochar. Moreover, the biochar enriched the compost with phosphorus and potassium while preserving nitrogen content. Additionally, increasing the amount of compost in growing media led to a significant rise in chlorophyll content in the leaves of garden cress and spinach during a 30-day growth period. Taken together, the findings suggest that the compost derived from spent coffee grounds and biochar is well-suited for enhancing the growth of garden cress and spinach, providing effective nutrients for these leafy vegetables, and facilitating soil–plant–atmosphere nutrient cycling. It is expected that composting spent coffee grounds with biochar could produce nutrient-rich biofertilizer in line with a circular economy strategy, offering a high-quality organic fertilizer suitable for the cultivation of leafy vegetables in agriculture. Returning waste organic resources to the soil can help build up carbon, reduce reliance on chemical inputs, and contribute to farmer welfare while minimizing waste accumulation and protecting the environment.

Author Contributions: Conceptualization, S.O. and A.A.; methodology, H.O., N.O. and R.K.; software and design, S.O.; validation S.O., H.O. and N.O.; formal analysis H.O. and N.O.; literature analysis, H.O., N.O., A.A., R.K. and S.O.E.; resources, H.O., A.A. and S.O.; data curation, S.O. and A.A.; writing—original draft preparative, S.O., N.O., R.K. and S.O.E.; writing—review and editing, R.K., A.A. and S.O.E.; visualization, H.O.; supervision, S.O. and S.O.E.; project administration, S.O. All authors have read and agreed to the published version of the manuscript.

Funding: This research received no external funding.

Institutional Review Board Statement: Not applicable.

Informed Consent Statement: Not applicable.

Data Availability Statement: All data are reported in the manuscript.

Conflicts of Interest: The authors declare no conflicts of interest.

References

1. Pehlivan, U.; Ozdemir, S.; Ozer, H.; Dede, O.H. Fuel properties and incineration behavior of poultry litter blended with sweet sorghum bagasse and pyrolysis oil. *J. Environ. Manag.* **2024**, *357*, 120844. [CrossRef] [PubMed]
2. Tutunchian, S.; Altınbaş, M. Assessment of an appropriate integrated waste management plan targeting the Circular Economy based on the LCA method. *J. Mater. Cycles Waste Manag.* **2023**, *25*, 456–478. [CrossRef]
3. Gebreeyessus, G.D. Towards the sustainable and circular bioeconomy: Insights on spent coffee grounds valorization. *Sci. Total Environ.* **2022**, *833*, 155113. [CrossRef] [PubMed]
4. Ozdemir, S.; Wynn, M.; Metin, B. Cybersecurity and country of origin: Towards a new framework for assessing digital product domesticity. *Sustainability* **2023**, *15*, 87. [CrossRef]
5. Picca, G.; Goñi-Urtiaga, A.; Gomez-Ruano, C.; Plaza, C.; Panettieri, M. Suitability of co-composted biochar with spent coffee grounds substrate for tomato (*Solanum lycopersicum*) fruiting stage. *Horticulturae* **2023**, *9*, 89. [CrossRef]
6. Dimitrijević, S.; Milić, M.; Buntić, A.; Dimitrijević-Branković, S.; Filipović, V.; Popović, V.; Salamon, I. Spent coffee grounds, plant growth promoting bacteria, and medicinal plant waste: The biofertilizing effect of high-value compost. *Sustainability* **2024**, *16*, 1632. [CrossRef]
7. Zengin, G.; Sinan, K.I.; Mahomoodally, M.F.; Angeloni, S.; Mustafa, A.M.; Vittori, S.; Maggi, F.; Caprioli, G. Chemical composition, antioxidant and enzyme inhibitory properties of different extracts obtained from spent coffee ground and coffee silverskin. *Foods* **2020**, *9*, 713. [CrossRef]
8. El-Sayed, I.M.; Salama, W.H.; Badr, M.A. Nitrogen and spent coffee ground for enhancing nutritional, morphological, flowering and antioxidant properties of Chrysanthemum (*Chrysanthemum morfolium* Ramat). *Plant Physiol. Biochem.* **2024**, *207*, 108406. [CrossRef]
9. Turp, G.A.; Ozdemir, S.; Yetilmezsoy, K.; Oz, N.; Elkamel, A. Role of vermicomposting microorganisms in the conversion of biomass ash to bio-based fertilizers. *Sustainability* **2023**, *15*, 8984. [CrossRef]
10. Guo, X.X.; Liu, H.T.; Zhang, J. The role of biochar in organic waste composting and soil improvement: A review. *Waste Manag.* **2020**, *102*, 884–899. [CrossRef]
11. Cervera-Mata, A.; Delgado, G.; Fernández-Arteaga, A.; Fornasier, F.; Mondini, C. Spent coffee grounds by-products and their influence on soil C-N dynamics. *J. Environ. Manag.* **2022**, *302*, 114075. [CrossRef] [PubMed]
12. Kul, R.; Yıldırım, E.; Ekinci, M.; Turan, M.; Ercisli, S. Effect of biochar and process water derived from the co-processed sewage sludge and food waste on garden cress' growth and quality. *Sustainability* **2022**, *14*, 16652. [CrossRef]
13. Bhunia, S.; Bhowmik, A.; Mallick, R.; Debsarcar, A.; Mukherjee, J. Application of recycled slaughterhouse wastes as an organic fertilizer for successive cultivations of bell pepper and amaranth. *Scientia Hort.* **2021**, *280*, 109927. [CrossRef]
14. Wang, H.; Su, Z.; Ren, S.; Zhang, P.; Li, H.; Guo, X.; Liu, L. Combined use of biochar and microbial agents can promote lignocellulosic degradation microbial community optimization during composting of submerged plants. *Fermentation* **2024**, *10*, 70. [CrossRef]
15. Dede, C.; Ozer, H.; Dede, O.H.; Celebi, A.; Ozdemir, S. Recycling nutrient-rich municipal wastes into ready-to-use potting soil: An approach for the sustainable resource circularity with inorganic porous materials. *Horticulturae* **2023**, *9*, 203. [CrossRef]

16. Ozdemir, S.; Ozdemir, S.; Ozer, H.; Yetilmezsoy, K. A techno-sustainable bio-waste management strategy for closing chickpea yield gap. *Waste Manag.* **2021**, *119*, 356–364. [CrossRef]
17. Khan, N.; Clark, I.; Sánchez-Monedero, M.A.; Shea, S.; Meier, S.; Bolan, N. Maturity indices in co-composting of chicken manure and sawdust with biochar. *Bioresour. Technol.* **2014**, *168*, 245–251. [CrossRef]
18. Ozdemir, S.; Dede, G.; Dede, O.H.; Turp, S.M. Composting of sewage sludge with mole cricket: Stability, maturity and sanitation aspects. *Int. J. Environ. Sci. Technol.* **2019**, *16*, 5827–5834. [CrossRef]
19. Santos, C.; Fonseca, J.; Aires, A.; Coutinho, J.; Trindade, H. Effect of different rates of spent coffee grounds (SCG) on composting process, gaseous emissions and quality of end-product. *Waste Manag.* **2017**, *59*, 37–47. [CrossRef]
20. Hagemann, N.; Subdiaga, E.; Orsetti, S.; de la Rosa, J.M.; Knicker, H.; Schmidt, H.P.; Kappler, A.; Behrens, S. Effect of biochar amendment on compost organic matter composition following aerobic composting of manure. *Sci. Total Environ.* **2018**, *613*, 20–29. [CrossRef]
21. Seninde, D.R.; Chambers IV, E.; Chambers, D. Determining the impact of roasting degree, coffee to water ratio and brewing method on the sensory characteristics of cold brew Ugandan coffee. *Food Res. Int.* **2020**, *137*, 109667. [CrossRef] [PubMed]
22. Ince, O.; Ozbayram, E.G.; Akyol, Ç.; Erdem, E.I.; Gunel, G.; Ince, B. Bacterial succession in the thermophilic phase of composting of anaerobic digestates. *Waste Biomass Valor.* **2020**, *11*, 841–849. [CrossRef]
23. Duan, Y.; Awasthi, M.K.; Yang, J.; Tian, Y.; Li, H.; Cao, S.; Ravindran, B. Bacterial community dynamics and co-occurrence network patterns during different stages of biochar-driven composting. *Bioresour. Technol.* **2023**, *384*, 129358. [CrossRef] [PubMed]
24. Ning, Y.; Liu, Y.; Guo, H.; Wang, X.; Yang, Y.; Zhou, D. Effect of the lignocellulolytic Psychrotroph Lelliottia sp. on bacterial community succession in corn straw compost. *Environ. Sci. Pollut. Res.* **2023**, *30*, 66346–66358. [CrossRef] [PubMed]
25. Wang, C.; Wu, M.; Peng, C.; Yan, F.; Jia, Y.; Li, X.; Qiu, Z. Bacterial dynamics and functions driven by a novel microbial agent to promote kitchen waste composting and reduce environmental burden. *J. Cleaner Prod.* **2022**, *337*, 130491. [CrossRef]
26. Zhao, N.; Liu, Z.; Yu, T.; Yan, F. Spent coffee grounds: Present and future of an environmentally friendly applications on industries-A review. *Trends Food Sci. Technol.* **2023**, *143*, 104312. [CrossRef]
27. Santhanarajan, A.E.; Han, Y.H.; Koh, S.C. The efficacy of functional composts manufactured using spent coffee ground, rice bran, biochar, and functional microorganisms. *Appl. Sci.* **2021**, *11*, 7703. [CrossRef]
28. Mironov, V.; Vanteeva, A.; Sokolova, D.; Merkel, A.; Nikolaev, Y. Microbiota dynamics of mechanically separated organic fraction of municipal solid waste during composting. *Microorganisms* **2021**, *9*, 1877. [CrossRef]
29. Aguilar-Paredes, A.; Valdés, G.; Araneda, N.; Valdebenito, E.; Hansen, F.; Nuti, M. Microbial community in the composting process and its positive impact on the soil biota in sustainable agriculture. *Agronomy* **2023**, *13*, 542. [CrossRef]
30. Aslam, S.; Nazir, A. Valorizing combustible and compostable fractions of municipal solid waste to biochar and compost as an alternative to chemical fertilizer for improving soil health and sunflower yield. *Agronomy* **2024**, *14*, 1449. [CrossRef]
31. Mierzwa-Hersztek, M.; Gondek, K.; Baran, A. Effect of poultry litter biochar on soil enzymatic activity, ecotoxicity and plant growth. *Appl. Soil. Ecol.* **2016**, *105*, 144–150. [CrossRef]
32. Ali, M.; Ahmed, O.H.; Jalloh, M.B.; Primus, W.C.; Musah, A.A.; Ng, J.F. Co-composted chicken litter biochar increases soil nutrient availability and yield of *Oryza sativa* L. *Land* **2023**, *12*, 233. [CrossRef]
33. Cicek, N.; Tuccar, M.; Yucedag, C.; Cetin, M. Exploring different organic manures in the production of quality basil seedlings. *Environ. Sci. Pollut. Res.* **2023**, *30*, 4104–4110. [CrossRef] [PubMed]
34. Titova, J.; Baltrėnaitė, E. Physical and chemical properties of biochar produced from sewage sludge compost and plants biomass, fertilized with that compost, important for soil improvement. *Waste Biomass Valor.* **2021**, *12*, 3781–3800. [CrossRef]
35. Caliskan, S.; Ozok, N.; Makineci, E. Utilization of spent coffee grounds as media for stone pine (*Pinus pinea*) seedlings. *J. Soil. Sci. Plant Nutr.* **2020**, *20*, 2014–2024. [CrossRef]
36. Gunes, A.Y.D.I.N.; Inal, A.; Sahin, O.; Taskin, M.B.; Atakol, O.; Yilmaz, N. Variations in mineral element concentrations of poultry manure biochar obtained at different pyrolysis temperatures, and their effects on crop growth and mineral nutrition. *Soil. Use Manag.* **2015**, *31*, 429–437. [CrossRef]

life

MDPI

Article

Rootstock Genotype Dictates Phosphorus Deficiency Tolerance and Transcriptional Plasticity in Grafted *Camellia oleifera* Plants

Zhihua Ren [1,2], Juan Liu [1,*], Jin Zeng [3], Li Cheng [1], Huiyun Liu [1], Yunyu Zhang [4], Qinhua Cheng [5], Wenjuan Su [6], Huaiyuan Wu [4] and Dongnan Hu [2,*]

1 College Forestry, Jiangxi Agricultural University, Nanchang 330045, China; zhihuaren2023@163.com (Z.R.); suosuodecl@163.com (L.C.); huiyunliu@foxmail.com (H.L.)
2 Jiangxi Key Laboratory of Subtropical Forest Resources Cultivation, Nanchang 330045, China
3 School of Mechanical Engineering, Wuhan Polytechnic University, Wuhan 420023, China; zengjin0819@163.com
4 Xinyu Forestry Development Service Center, Xinyu 338099, China; yyzh1949@163.com (Y.Z.); ericwhy1993@163.com (H.W.)
5 Gannan Arboretum, Ganzhou 341299, China; qinhua19990315@163.com
6 College of Forestry and Grassland, Nanjing Forestry University, Nanjing 210037, China; suwenj0126@163.com
* Correspondence: liujuan@jxau.edu.cn (J.L.); dnhu98@jxau.edu.cn (D.H.)

Abstract: Rootstock choice offers a powerful lever for tailoring economically important trees to adverse environments. *Camellia oleifera* Abel., a premier oil-producing species cultivated widely on red-soil hills, suffers large yield losses under chronic phosphorus deficiency. We grafted a single elite scion (CL4) onto three contrasting rootstocks (CL4, CL3, CL53) and monitored growth and root transcriptomes for 1.5 years under adequate (1 mM) or limiting (0 mM) P supply. Under low-P stress, the rootstock identity reshaped the root architecture: CL4/CL3 produced the longest, most extensive network, increasing the total root length by 49.7%, the surface area by 52.9%, and the volume by 42.6% relative to the control, whereas leaf morphology responded solely to P supply, not to the graft combination. CL4/CL3 also accumulated up to more than 17.5% of root biomass and 28.25% of whole-plant biomass than any other combination. Physiologically, CL4/CL3 acted as an aggressive P miner, accumulating 67.8% more P in its roots than the self-grafted control under P limitation, while CL4/CL4 maximized the internal P use efficiency, showing a 44.74% higher root P use efficiency than CL4/CL53—two contrasting yet effective strategies for coping with low-P stress. Transcriptomic profiling uncovered 1733 DEGs in the CL4/CL3 and 2585 in the CL4/CL4 roots, with 150 and 255 uniquely co-expressed genes, respectively. CL4/CL3 up-regulated organic-acid and phenylpropanoid pathways; CL4/CL4 activated defense and phosphate transport networks. qRT-PCR of six genes confirmed that CL4/CL3 mounted a stronger low-P response via MAPK, hormonal, and lipid–metabolic signaling. These results provide a mechanistic framework for rootstock-mediated P efficiency and establish a foundation for the molecular breeding of *C. oleifera* under nutrient-limited conditions.

Keywords: low-P stress; *Camellia oleifera*; rootstocks; transcriptomics

1. Introduction

Phosphorus (P) is an essential element for plants, significantly influencing their growth and development [1]. It plays a vital role in various plant functions, such as metabolic regulation, energy transfer, and protein activation [2–4]. While P is mainly present in natural soil, it is mostly in the form of organophosphorus compounds, leading to a low concentration of inorganic phosphate (Pi) available for plant uptake [5]. Over the past century, the excessive use of P fertilizers has been driven by the need to meet food demands and maintain high crop yields but has resulted in increased costs and negative environmental impacts [6]. Moreover, the depletion of rock phosphate, a key component of these fertilizers, is a growing concern, with reserves projected to be exhausted within the next 50–100 years [7,8].

Native plant species have evolved unique strategies to adapt to low-Pi availability in soil [9]. Progress has been made in enhancing plant P acquisition efficiency through root foraging and phosphorus use efficiency (PUE) through P remobilization strategies [10,11]. For example, the presence of a growing number of root hairs and a relative high root-to-shoot ratio have been considered as symbiotic traits that contribute to high P acquisition efficiency [12,13]. Studies have shown that by increasing the length of the root system, the number of lateral roots, and the branching angle of the root system, plants increase the contact area between the root system and the soil, thereby increasing P absorption [11,12]. For instance, cotton plants respond to low-P stress by enhancing P absorption through increased lateral roots, branching density, and root hair length [14]. Furthermore, the response to Pi starvation is consistently regulated by gene regulatory networks, which involves numerous Pi transporters and Pi starvation-induced genes [15]. Many transcription factors, signal molecules, and other upstream regulators could be activated by Pi starvation. For example, the transcript factor WRKY has been demonstrated to participate in the regulation of Pi acquisition under low-Pi condition [16].

Grafting is a cornerstone horticultural technique that accelerates clonal propagation while preserving elite traits and stabilizing heterosis. By reshaping root architecture and physiology, it also enhances tolerance to environmental stresses, including water stress and nutrient deficiency [17,18]. Given that rootstocks are responsible for nutrient absorption, grafted plants can take up more ions and water compared to self-rooted plants and transport these to the scion. Utilizing the grafting technique to select a highly tolerant root system as a rootstock under nutrient deficiency can enhance nutrient absorption and/or utilization in plants. For example, Adam & Ulas (2023) reported that the tomato genotypes HelenaF1 and ALT exhibit exceptional rootstock potential, manifested through coordinated morphological, physiological, and biochemical traits that confer pronounced resilience under low-nitrogen conditions [19]. In citrus, the FA-5 rootstock confers superior drought tolerance by enhancing the osmotic adjustment capacity, thereby markedly improving the performance of grafted trees under water deficit conditions [20]. Yet, the rootstock-mediated mechanisms that orchestrate P acquisition, remobilization, and signaling under chronic P limitation—particularly in long-lived woody species—remain largely unexplored.

Camellia oleifera Abel. is an essential woody oil-producing tree species endemic to China. The tea oil extracted from the seeds of this species is highly valued for human health and is favored by consumers as a high-quality edible oil. With the growing oil demand, *C. oleifera* planting area has been continuously expanding in recent decades. It has been reported that the cultivated areas have reached 4.5 million ha in 2022 and will be 6 million ha in 2025. This species is naturally grown and widely planted in the acidic red soil regions of Southern China, which are characterized by a notable P deficiency. The shortage of Pi in these areas has significantly hampered the development of the *C. oleifera* industry [16]. In practice, the plants of *C. oleifera* are often propagated by sprout anvil grafting [21]. Many studies have focused on the variability in P among different scion varieties. For example, when the vigorous variety of the 'Gan 8' scion was grafted with the rootstock of 'Ganwu 2', it showed better performance in shoot P accumulation and root PUE under low-P stress [22]. A comprehensive assessment of seedling from twelve *C. oleifera* half-sib families revealed that the plants from 'Changlin 4' (CL4) half-sib families, which had higher root/shoot ratio, root area, and PUE, were categorized as P-efficient. On the other hand, the plants from 'Changlin 3' (CL3) half-sib families performed poorly in these features and were categorized as P-inefficient. However, the rootstock-mediated effect on low-P stress in *C. oleifera* has yet to be investigated. It is still unknown whether the tolerance of the grafted seedling to a low P supply would vary when different types of P-efficient plants are used as rootstocks for the same scion.

In this study, we investigated the response of grafted *C. oleifera* combinations, with the same commercial scion variety grafted onto three different rootstocks under P deficiency and control conditions (0 mM and 1 mM). The research focused on the following aspects: (1) the influence of rootstock and P availability on the morphology and biomass of the grafted plants; (2) the impact of rootstock and P availability on P accumulation and PUE of the grafted plants; (3) the effect of rootstock and P availability on unigenes and their underlying signaling pathways.

2. Materials and Methods

2.1. Plant Materials and Growth Conditions

The experiment was conducted at the Science Park of Jiangxi Agricultural University (115°49′ E, 28°45′ N), in Nanchang, Jiangxi Province, China. The locale is characterized by a subtropical monsoon climate, experiencing 147–157 precipitation days annually, with the annual rainfall ranging between 1600 and 1700 mm and an average annual temperature fluctuating between 17 °C and 17.7 °C. Building on our earlier findings [23], we employed three semi-sibling rootstocks that differ markedly in phosphorus use efficiency (PUE): CL3 (low PUE), CL4 (high PUE), and 'Changlin 53' (CL53; moderate–high PUE). The scion chosen for all grafting combinations was the 'Changlin 4' (CL4) cultivar. Consequently, we produced three grafted assemblies—CL4/CL3, CL4/CL4, and CL4/CL53 (scion/rootstock)—with the self-grafted CL4/CL4 acting as the internal benchmark, and CL4/CL3 and CL4/CL53 as the experimental grafts. This single-scion design stripped away scion variability and directly exposed each rootstock's influence on phosphorus acquisition, echoing the commercial practice of pairing uniform elite clones with diverse rootstocks to tailor orchards to specific stresses.

In December 2018, *Camellia* seeds underwent cold stratification in sand, a process critical for germination. By May 2019, these germinated seeds were selected for anvil grafting. The grafting methodology was described by [22]. The germinated *C. oleifera* seeds were trimmed 2–3 cm above the seed, and the trimmed roots were extracted as rootstocks. Branches and leaves, selecting one leaf per scion, were collected from the *C. oleifera* germplasm resource nursery. The wedge-shaped tip of each scion branch was inserted into a V-shaped cleft in the root, which had been cut vertically for 1 cm, and was subsequently wrapped and then placed on the seedbed and cultivated to form a grafted plant for experimental treatment. A grafting substrate, a mixture of peat, perlite, and vermiculite in a 3:1:1 ratio (available P: 14.95 ± 0.26 mg/kg, available potassium: 464.39 ± 3.34 mg/kg, hydrolyzable nitrogen: 419.01 ± 6.77 mg/kg), was utilized, resulting in a high survival rate among all grafted plants, without significant variance in compatibility among the three graft combinations. In January 2021, a critical transplantation stage was initiated, transferring the plants into pots with a river sand substrate (available P: 1.06 ± 0.14 mg/kg, available potassium: 27.19 ± 0.33 mg/kg, hydrolyzable nitrogen: 5.36 ± 1.23 mg/kg). Commencing in June 2020, healthy and uniform grafted plants were subjected to experimental treatments. Prior trials showed that *C. oleifera* survives in P-impoverished red soils; 0 mM P has therefore been our validated threshold for P-stress studies. Therefore, P regimes were set as 1 mM (control) and 0 mM (deficiency). KH_2PO_4 was used as the P source. Since June 2020, the grafted plants were irrigated monthly with 100 mL of 1/2 Hoagland nutrient solution (no P included). Standard horticultural practice was maintained by applying 100 mL of tap water weekly and adjusting the amount according to weather, increasing irrigation during hot periods and reducing it when it rained. To disentangle acute from chronic responses to phosphorus limitation, we implemented a staggered-harvest time series (0, 30, 90, and 180 d). In November 2021, all grafted plants were collected and immediately dissected into distinct tissues for downstream analyses. Each treatment combination was represented by three independent biological replicates to guarantee the robustness and reproducibility of our findings.

2.2. Determination of the Morphology

Leaf morphometrics, including leaf area, length, width, and perimeter, were quantitatively assessed using a specialized leaf area measuring instrument of the handheld laser leaf area meter CI-203-CID (CID Bio-Science, Camas, WA, USA). Subsequently, the leaf shape index was calculated as the ratio of leaf length to leaf width. Three measurement rounds were taken on every plant ($n = 3$ per treatment, all plants measured, no random selection).

The entire root systems of the *C. oleifera* grafted plants were carefully excavated and cleansed for root analysis. The roots were then meticulously severed at the branching nodes and arranged in a transparent plastic container for imaging. Utilizing a scanner Expression 10000XL3.49 (Epson America, Los Alamitos, CA, USA), high-resolution scans of the entire root system were obtained. Advanced image analysis was performed using the Win Rhizo Pro2012b (Regent Instruments Inc., Quebec, QC, USA) software, enabling the precise quantification of total root length, surface area, and volume.

2.3. Determination of Biomass and P Concentration

All grafted plants were meticulously harvested. The roots and shoots were segregated and individually encased in brown paper envelopes. At each of the four harvest time points, triplicate samples were taken from three different grafted plants per treatment and initially placed in an oven set at 105 °C for two hours, followed by a drying phase at 65 °C until a constant weight was achieved. Each tissue's mass was precisely measured using an electronic balance with a sensitivity of 0.01 g. The root-to-shoot ratio and total dry weight were subsequently calculated. The root-to-shoot ratio, expressed in g/g, was determined as the root weight divided by the shoot weight. The total dry weight, in grams, was computed as the sum of the dry weights of the root and shoot.

For P content analysis, three replicated samples per treatment were analyzed using the molybdenum–antimony colorimetric method, with measurements conducted via a GENESYS 180 UV-Vis spectrophotometer (Thermo Fisher Scientific, Waltham, MA, USA). Phosphorus use efficiency (PUE) was estimated based on the total P accumulation by the plant. P accumulation, measured in mg, was calculated as the product of P concentration (mg/g) and dry weight (g). PUE, denoted in g/mg, was determined as the dry weight divided by the total P accumulation.

2.4. RNA Extraction, Illumina Sequencing, and Data Analysis

In order to elucidate the P deficiency rootstock-mediated adaptability in *C. oleifera*, we focused on the rootstocks CL3 and CL4 because of their differential performance under the P deficiency condition and further explored the underlying molecular mechanisms using root transcriptome sequencing. In this study, the roots of the CL4/CL3 (scion/rootstock) combination in the control and P deficiency treatment groups were designated as R34P1 and R34P0, respectively. Similarly, the roots of the CL4/CL4 combination under control and low-P conditions were labeled as R44P1 and R44P0, respectively. Each treatment had three replicates.

Total root RNA extraction was performed using TRIzol® reagent (Thermo Fisher Scientific, Waltham, MA, USA) according to the manufacturer's protocol. RNA integrity and quantity were assessed using a 5300 Bioanalyzer Agilent (Agilent Technologies, Santa Clara, CA, USA) and an ND-2000 spectrophotometer (Thermo Fisher Scientific, Waltham, MA, USA), respectively. RNA samples with high quality were selected for the construction of the sequencing library.

The RNA-seq transcriptome library was prepared using Illumina® Stranded mRNA Prep and a ligation method, starting with 1 μg of total RNA. Subsequently, double-stranded cDNA synthesis was performed using the SuperScript double-stranded cDNA synthesis kit (Thermo Fisher Scientific, Waltham, MA, USA) with random hexamer primers from Illumina. Post-quantification with Qubit 4.0 (Thermo Fisher Scientific, Waltham, MA, USA), the paired-end RNA-seq sequencing library was processed on the NovaSeq 6000 sequencer (Illumina, San Diego, CA, USA), generating reads of 2 × 150 bp. RNA purification, reverse transcription, library construction, and sequencing were completed by Shanghai Majorbio Bio-pharm Biotechnology Co., Ltd., Shanghai, China.

In this study, raw paired-end reads were meticulously trimmed and subjected to quality control using fastp with its default parameters. Clean reads were independently aligned to the reference genome in orientation mode utilizing HISAT2 (http://ccb.jhu.edu/software/hisat2/index.shtml, accessed on 5 November 2022). Subsequently, StringTie was employed in a reference-based approach to assemble the mapped reads of each sample. Differentially Expressed Genes (DEGs) between two distinct samples were identified by calculating the expression level of each transcript using the Transcripts Per Million reads (TPM) method, quantifying gene abundances. Further, functional enrichment analysis, including Gene Ontology (GO) and Kyoto Encyclopedia of Genes and Genomes (KEGG) pathways, identified DEGs significantly enriched in GO terms and metabolic pathways, with a Bonferroni-corrected p-value of $\leq$0.05, in comparison to the entire transcriptome background. GO functional enrichment and KEGG pathway analyses were executed.

2.5. Quantitative Real-Time PCR Analysis

Several candidate genes were selected to validate the transcript expression using quantitative real-time PCR (qRT-PCR). RNA extraction was carried out using RNAprep Pure DP441 (Tiangen, Beijing, China). The reverse transcription of RNA to synthesize cDNA was performed according to the iScript™ cDNA Synthesis Kit protocol (Bio-Rad, Hercules, CA, USA), utilizing a PCR instrument 846-x 070-301 (Bio-Rad, Hercules, CA, USA) set to a specific temperature and time program (25 °C for 5 min, 46 °C for 20 min, 95 °C for 1 min, and holding at 4 °C).

Primers for these genes were designed based on the sequencing results, using the online tool available at Primer3 (https://www.primer3plus.com/, accessed on 12 October 2023) online and synthesized by Shanghai Majorbio Bio-pharm Biotechnology Co., Ltd., Shanghai, China (Table 1). EF1a1 was chosen as the internal reference gene. Real-time quantitative detection was conducted using iTaq™ Universal SYBR® Green Supermix 1725121 (Bio-Rad, Hercules, CA, USA) in a CFX Connect™ Real-Time System 788BR07708 (Thermo Fisher Scientific, Waltham, MA, USA), following a temperature and time program of 95 °C for 2 min, 95 °C for 5 s, 60 °C for 30 s, for 40 cycles, with a final to obtain a melt curve (95 °C for 5 s, 65 °C for 5 s, and 95 °C for 5 s). The relative gene expression levels were quantified using the $2^{-\Delta\Delta Ct}$ method.

2.6. Statistical Analysis

Two-factor analysis of variance (ANOVA) was employed to investigate the effects of rootstock varieties and P availability, as well as their interaction effects, in the different grafted plants. One-way ANOVA was employed to assess differences between the control and the P deficiency conditions using Student's *t*-test at the 5% significance level.

Table 1. Primer Information.

Name	Gene id	Description	Primer Sequences
MAPK	augustus_masked-HiC_scaffold_13-processed-gene-1462.3	MAPK signaling pathway	F: CACTCGCCACAAACAATCCC R: TACGCTTGTGGGTTGTCGAA
CBs	snap_masked-HiC_scaffold_6-processed-gene-823.4	Carotenoid biosynthesis	F: CTCAGAGCGTGATGGGGATC R: ATTCTTGTTGAGCCGAGGCA
ELM	maker-HiC_scaffold_14-snap-gene-732.64	Ether lipid metabolism	F: TACCCATGGTGGTTGCCTTC R: ACGGGCCTGACTAATTGCAT
FBs	augustus_masked-HiC_scaffold_10-processed-gene-1746.20	Fatty acid biosynthesis	F: GTCCCACTCCGACATTCTCC R: TGGTTAAGTCGGTTGGAGGC
PPI	augustus_masked-HiC_scaffold_1-processed-gene-131.29	Plant–pathogen interaction	F: CCAGTGGCGGAGTCCAAATA R: GGGCGGGTCATGATGTAGAG
PHST	snap_masked-HiC_scaffold_14-processed-gene-167.29	Plant hormone signal transduction	F: GTGCCCTCTCAAATGGTGGA R: GGCCTAAGCTAGCACTACCG

3. Results

3.1. Effects of Different Rootstocks on the Morphology of C. oleifera Grafted Plants

In this study, the morphological characters of leaves and roots were assessed among the *C. oleifera* plants grafted with CL4, CL3, and CL53 rootstocks under control and P deficiency conditions. The results showed that P availability could affect significantly the leaf area and leaf circumference of the *C. oleifera* grafted plants (Figure 1), but no significant difference in leaf morphologies was found among grafted plants with different rootstock genotypes.

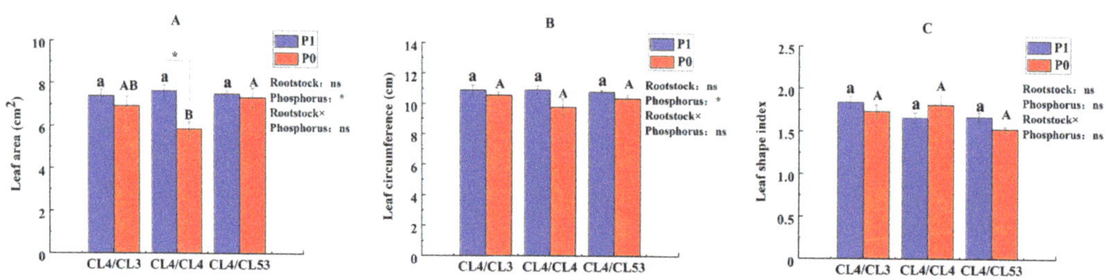

Figure 1. Effects of the rootstock on leaf morphology of the grafted plants under control and low-P conditions. (**A**) Leaf area, (**B**) leaf circumference, and (**C**) leaf shape index. Lowercase and uppercase letters on the bars indicate significant differences in grafting combinations between control and P deficiency supply treatments at the $p \leq 0.05$ level, respectively. (The statistical test used was one-way analysis of variance). The upper right of the figure shows the 'Interaction Effect Analysis' (the statistical test used was two-way analysis of variance); 'Rootstock', 'Phosphorus', and 'Rootstock × Phosphorus' indicate 'the impact of rootstock genotypes', 'the impact of phosphorus', and 'the interaction of rootstock genotypes and phosphorus', respectively. '*' indicate that the interactions are significant at the $p \leq 0.05$ level, respectively, and 'ns' indicate no significant difference.

The results of the root morphology analysis indicated that P availability did not affect the root characters. The rootstock genotypes significantly affected the root length and total root surface area but did not influence the total root volume (Figure 2). Among the *C. oleifera* plants grafted with different rootstocks, CL4/CL3 boosted the total root length by 49.7%, the surface area by 52.9%, and the root volume by 42.6% relative to the control under P deficiency.

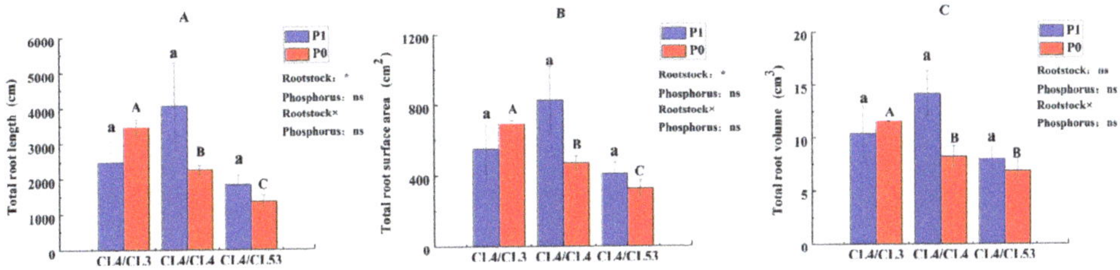

Figure 2. Root morphology of *C. oleifera* plants. (**A**) Total root length, (**B**) total root surface area, and (**C**) total root volume. Lowercase and uppercase letters on the bars indicate significant differences in grafting combinations between control and P deficiency supply treatments at the $p \leq 0.05$ level, respectively. (The statistical test used was one-way analysis of variance). The upper right of the figure shows the 'Interaction Effect Analysis' (the statistical test used was two-way analysis of variance); 'Rootstock', 'Phosphorus', and 'Rootstock × Phosphorus' indicate 'the impact of rootstock genotypes', 'the impact of phosphorus', and 'the interaction of rootstock genotypes and phosphorus', respectively. '*' indicate that the interactions are significant at the $p \leq 0.05$ level, respectively, and 'ns' indicate no significant difference.

3.2. Effect of the Rootstock on the Biomass of Grafted C. oleifera

The results of the biomass of *C. oleifera* indicated that root biomass, shoot biomass, whole-plant biomass, and root/shoot ratio in the grafted plants of three rootstocks were not affected by P availability and the rootstock genotypes (Figure 3). However, we found that different grafted plants had different performance under the P deficiency treatment. For example, the root biomass and the entire plant biomass of CL4/CL3 were much greater compared to those of the other grafted plants. Under P deficiency treatment, the root biomass of CL4/CL3 increased by 17.51% compared with that of the control and was 4.06% higher compared to that of CL4/CL4. The whole-plant biomass of CL4/CL3 under low P supply was improved by up to 28.25% compared to that of the control.

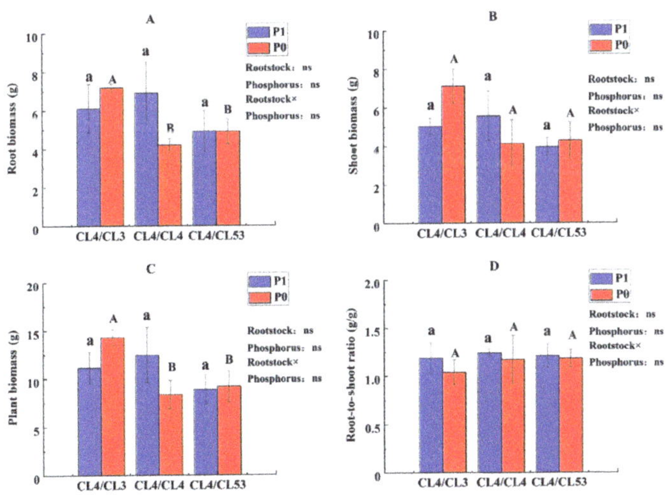

Figure 3. Biomass of each part of the grafted plants. (**A**) Root biomass, (**B**) shoot biomass, (**C**) whole-plant biomass, and (**D**) root/shoot ratio. Lowercase and uppercase letters on the bars indicate significant differences in grafting combinations between control and P deficiency supply treatments at the $p \leq 0.05$ level, respectively. (The statistical test used was one-way analysis of variance). The upper right of the figure shows the 'Interaction Effect Analysis' (the statistical test used was two-way figure

shows the 'Interaction Effect Analysis' (the statistical test used was two-way analysis of variance); 'Rootstock', 'Phosphorus', and 'Rootstock × Phosphorus' indicate 'the impact of rootstock genotypes', 'the impact of phosphorus', and 'the interaction of rootstock genotypes and phosphorus', respectively. 'ns' indicate no significant difference.

3.3. Effects of the Rootstock on P Content, P Accumulation, and PUE in the Grafted *C. oleifera*

We further investigated the P content, P accumulation, and PUE among the *C. oleifera* plants grafted with three rootstock genotypes in different tissues under control and P deficiency availability. The findings indicated that root P content, root P accumulation, and root PUE in the grafted plants were significantly affected by both rootstock genotypes and P availability (Figure 4). Meanwhile, whole-plant P accumulation and whole-plant PUE exhibited similar performance and were significantly affected by both rootstock × P, rootstock variety, and P availability. But shoot P content, shoot accumulation, and shoot PUE in the grafted plants were mainly affected by P availability.

Among the three graft combinations, CL4/CL3 achieved the greatest P accumulation in roots, shoots, and whole plants under P-limited conditions; for instance, root P accumulation was 67.8% higher than in the self-grafted CL4/CL4 control. On the other hand, CL4/CL4 exhibited modestly superior root P use efficiency—44.74% higher than that of CL4/CL53—and, under P deficiency, maintained the highest whole-plant PUE while accumulating less P in every tissue under both treatments. It seemed that CL4/CL3 and CL4/CL4 had different strategies to adapt to a P-deficient environment. The former showed high P-mining capacity by increasing root P absorption from the soil and accumulating P in the whole plant to cope with the P deficiency. The latter enhanced P use efficiency by optimizing the P application rate to meet the plant's P demand. Therefore, CL4/CL3 and CL4/CL4 were selected for further transcriptomic analysis under two P availability levels in this study to clarify the underlying mechanisms.

3.4. Transcriptomic Analysis for Different Grafted Plants Under Two Treatments

Based on the different performance on P content, P accumulation, and PUE under low P levels, the roots of the CL4/CL3 (scion/rootstock) and CL4/CL4 grafted combinations were chosen for further transcriptional sequencing analysis to investigate the response mechanisms of plants grafted with different rootstocks (Figures 5 and 6). PCA analysis of all samples revealed a strong correlation within each treatment and significant differences between the treatment groups, indicating sufficient reliability for further analysis (Figure 5A). The results of DGE analysis between control and P deficiency conditions showed that there were 1064 up-regulated genes and 669 down-regulated genes in the CL4/CL3 roots, while 1415 up-regulated and 1170 down-regulated genes were found in the CL4/CL4 roots (Figure 5B). The Venn diagram was used to present the presence of uniquely expressed genes in each of the four combinations. The results indicated that there were 150 co-expressed genes in the CL4/CL3 combination under the two different treatments, and 255 genes in the CL4/CL4 combination (Figure 5C).

The GO enrichment analysis revealed distinct functional specializations of root genes in the CL4/CL3 and CL4/CL4 graft combinations when investigating the DEGs of each combination under the two P levels (Figure 6). When compared to the DEGs of the CL4/CL3 roots under control and P deficiency conditions, gene functions primarily related to monocarboxylic acid biosynthesis, phenylpropane metabolism, secondary metabolite biosynthesis, phosphoenolpyruvate family amino acid catabolism, and cinnamic acid metabolism (Figure 6A). The DEGs in the CL4/CL4 roots under the two different treatments were primarily involved in pathways associated with defense reaction, hemicellulose metabolism, phosphate ion transport, phosphoenolpyruvate family amino acid catabolism, and cell wall macromolecular metabolism (Figure 6B).

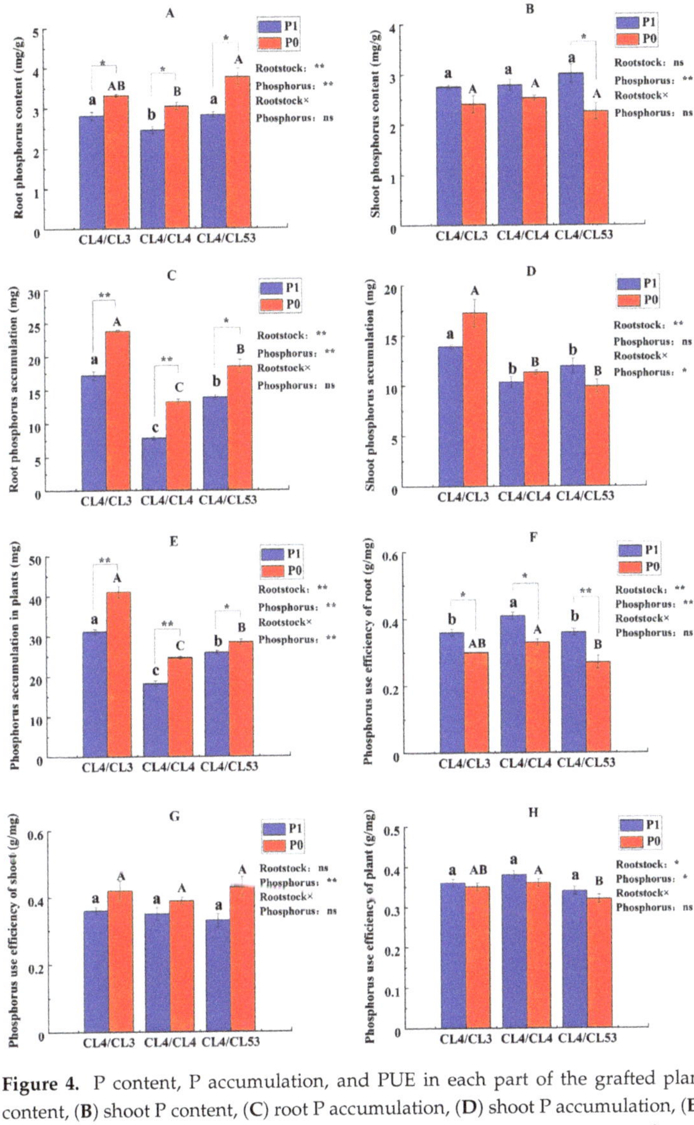

Figure 4. P content, P accumulation, and PUE in each part of the grafted plants. (**A**) Root P content, (**B**) shoot P content, (**C**) root P accumulation, (**D**) shoot P accumulation, (**E**) whole-plant P accumulation, (**F**) root PUE, (**G**) shoot PUE, and (**H**) plant PUE. Lowercase and uppercase letters on the bars indicate significant differences in grafting combinations between control and P deficiency supply treatments at the $p \leq 0.05$ level, respectively. (The statistical test used was one-way analysis of variance). '*' and '**' on the bars indicate a significant difference between control and low-P-treated plants at the $p \leq 0.05$ and $p \leq 0.01$ levels, respectively. (The statistical test used was Student's test). The upper right of the figure shows the 'Interaction Effect Analysis' (the statistical test used was two-way analysis of variance); 'Rootstock', 'Phosphorus', and 'Rootstock × Phosphorus' indicate 'the impact of rootstock genotypes', 'the impact of phosphorus', and 'the interaction of rootstock genotypes and phosphorus', respectively. '*' and '**' indicate that the interactions are significant at the $p \leq 0.05$ and $p \leq 0.01$ levels, respectively, and 'ns' indicate no significant difference.

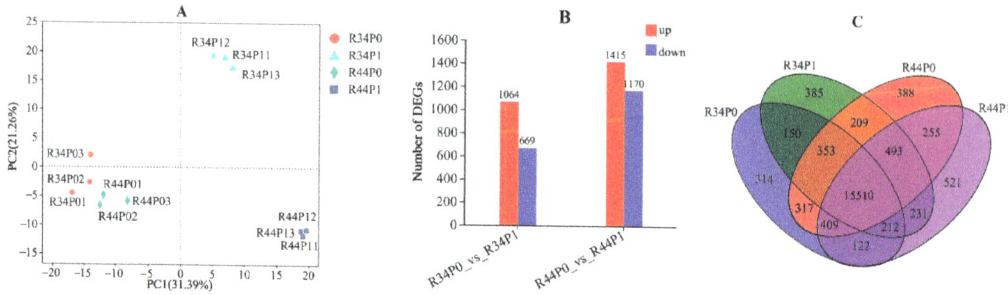

Figure 5. Comparative analysis of differential genes under control and low P conditions. (**A**) PCA analysis of the samples, (**B**) up-regulated and down-regulated DEGs, and (**C**) Venn diagram with the number of DEGs. 'R34P01' indicates 'R + 3 + 4 + P0 + 1', which represents 'root + rootstock (CL3 or CL4) + scion (CL4) + phosphorus level (P0 or P1) + repeat (1, 2 or 3)'.

KEGG analysis based on the DEGs in the CL4/CL3 roots under the two P levels showed a significant enrichment in phenylalanine metabolism, starch and sucrose metabolism, glutathione metabolism, pentose and glucuronate interconversions, plant hormone signal transduction, glycosphingolipid biosynthesis, and flavonoid biosynthesis. KEGG analysis of the CL4/CL4 roots under the two different treatments indicated a notably enrichment in phenylpropane biosynthesis, plant–pathogen interaction, plant hormone signaling, flavonoid biosynthesis, glutathione metabolism, zeatin biosynthesis, MAPK signaling pathway, and other metabolic pathways.

In order to verify the accuracy of the gene expression analysis, six genes based on the transcriptomic sequencing analysis were selected to conduct qRT-PCR under the two P treatments (Table 1). The results showed that the relative gene expression using RNA-Seq and qRT-PCR of five genes were matched (except for MAPK), indicating that the transcriptomic sequencing analysis was reliable (Figure 7). For example, the expression of MAPK was induced in the CL4/CL3 grafted combination under P deficiency, with an expression level 2.73 times higher than that in the control. In the CL4/CL4 grafted combination, the expression level of MAPK was 1.28 times higher than that in the control. The gene *CBs* in the grafted CL4/CL4 showed similar low expression levels under both treatments. But the expression level of this gene in CL4/CL3 under P deficiency was 2.26 times higher than that in the control. The lipid metabolism genes *ELM* and *FBs* are involved in P metabolism. The expression of *ELM* gene was induced by P deficiency, with similar performance in both grafted combinations. The expression of the *FBs* gene was repressed in the two grafted plants under P deficiency conditions compared to the control. The expression of PPI (a gene involved in plant–pathogen interaction) in both grafted plants was significantly induced in P deficiency conditions. And the expression level of this gene in CL4/CL3 was much higher than in CL4/CL4. The expression of PHST (a plant hormone signaling gene) was also induced under P deficiency conditions, being three times higher in CL4/CL3 than in CL4/CL4. This suggests that the P deficiency condition could induce the expression of hormone-related genes, Overall, P deficiency conditions stimulate intracellular signaling, leading to the secretion of hormones and metabolites like carotenoids, which then regulate lipid metabolism and facilitate the plant's response to P deficiency.

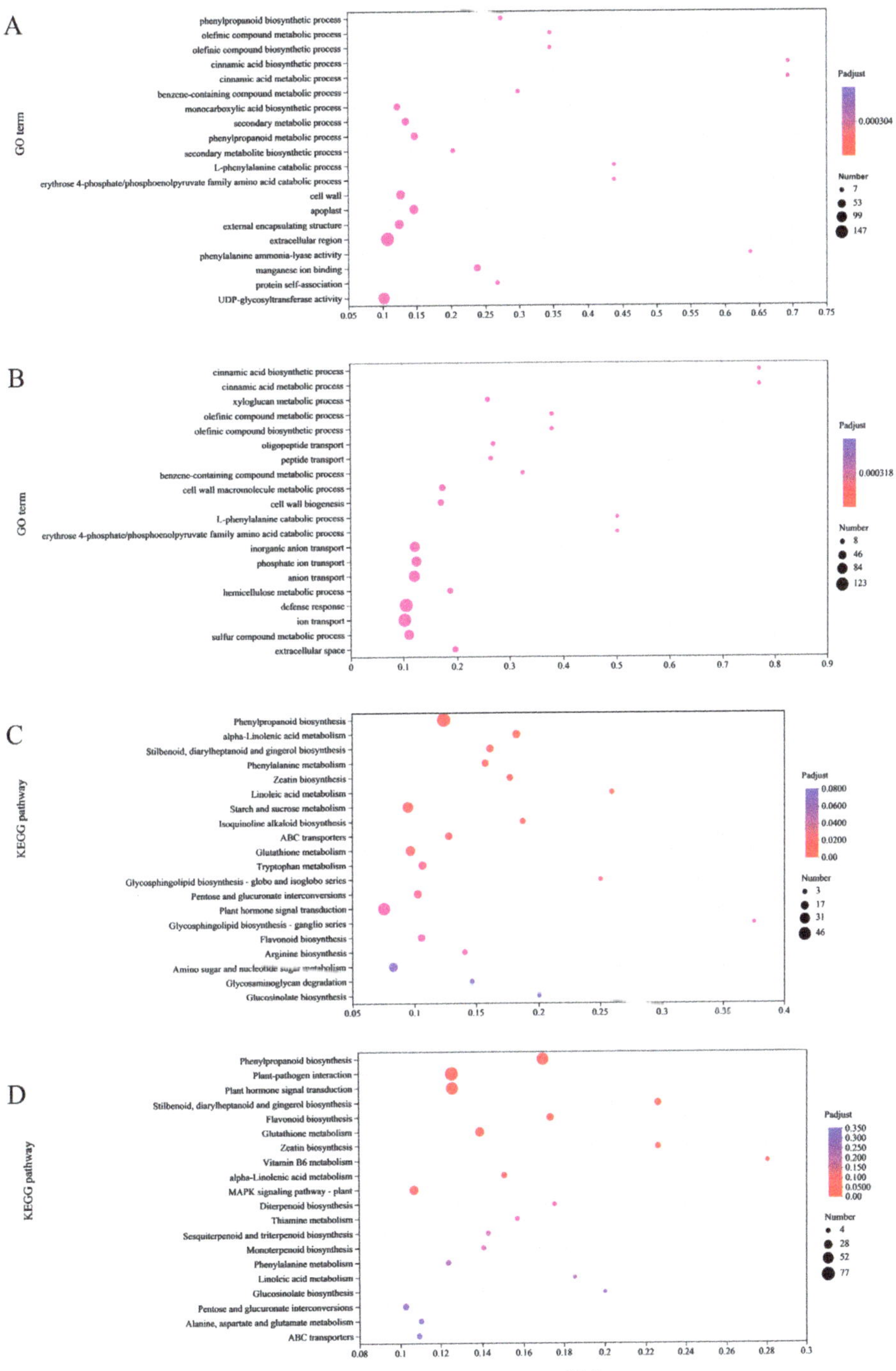

Figure 6. GO and KEGG enrichment analysis. (**A**) R34P1 vs. R34P0 GO enrichment analysis of DEGs, (**B**) R44P1 vs. R44P0 GO enrichment analysis of DEGs, (**C**) R34P1 vs. R34P0 KEGG enrichment analysis of DEGs, and (**D**) R44P1 vs. R44P0 KEGG enrichment analysis of DEGs.

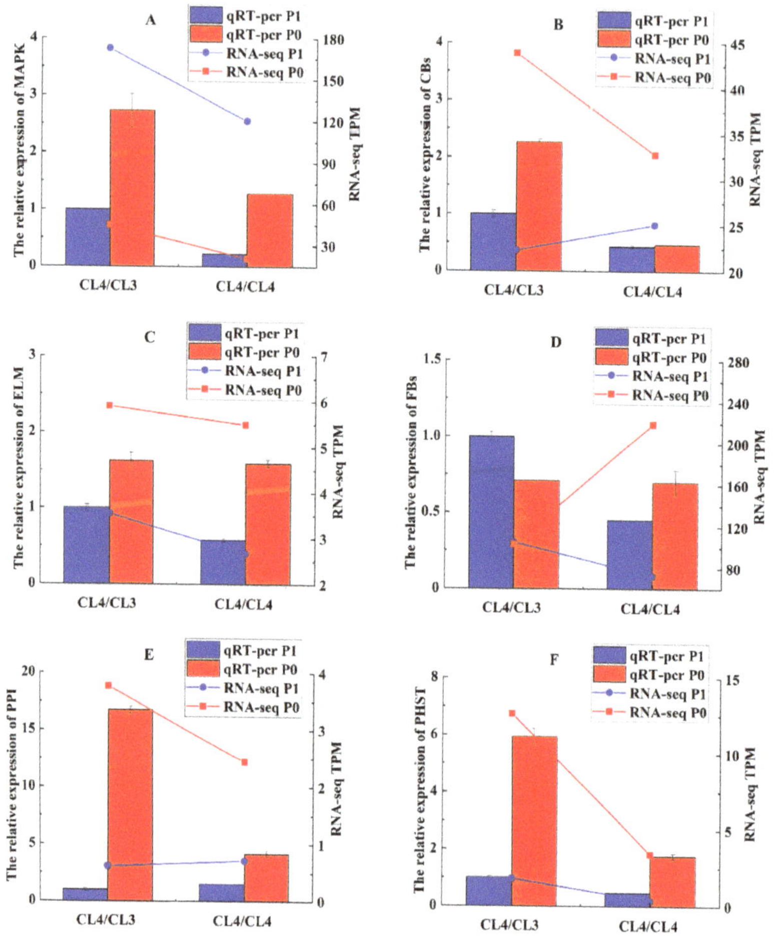

Figure 7. Real-time quantitative expression analysis and RNA-Seq validation analysis. (**A**) MAPK, (**B**) CBs, (**C**) ELM, (**D**) FBs, (**E**) PPI, and (**F**) PHST.

4. Discussion

Breeding highly nutrient-efficient varieties has been demonstrated as an effective way to address the nutrient shortage [24]. Grafting scion varieties maintains good bud characteristics while improving the plant's ability to adapt to environmental stress [25,26]. A growing body of research has begun to dissect how rootstocks modulate plant performance under phosphorus-limited conditions, with particular attention given to apple [27], peach [28], and grapevine [29]. In this study, we examined the effects of P deficiency on the same scion of *C. oleifera* grafted with different rootstocks. The results showed that the rootstocks of the *C. oleifera* grafted plants did not affect the leaf morphology but had a significant impact on the root length and total root surface area. Moreover, different grafted plants were found to have different root biomass, root P content, and PUE under P deficiency. The CL4/CL3 grafted combination performed better in nutrient absorption under P-limited conditions.

Phosphorus deficiency has been widely documented to impair growth and development across a broad range of crop species, including soybean [30] and cotton [31,32]. In this study, under P-limited conditions, the leaf area and leaf circumference were reduced apparently in three grafted combinations. This is because P deficiency could reduce stomatal opening, with less captured CO_2 and reduced triose phosphate, significantly influencing the recycling of NADPH and ATP, thus inhibiting plants' photosynthetic capacity and growth [33]. Moreover, rootstocks have been reported to alter scion morphology in some fruit trees, including apple [34,35], peach [36], and pear [37], where dwarfing or invigorating rootstocks can modulate leaf size, thickness, and specific leaf area via changes in endogenous hormone levels and hydraulic conductance. In this study, the three examined rootstock genotypes had little impact on the leaf morphologies of the grafted plants. Rootstocks would be more likely to modulate below-ground traits than above-ground leaf architecture under P limitation. In this study we observed that the three rootstocks scarcely altered scion leaf morphology, yet significantly changed root length and surface area. This divergence suggests that rootstock-mediated P acquisition operates primarily through root plasticity—enhancing soil exploration and P uptake—rather than through alterations in leaf size or shape.

Grafting is often considered to improve the nutrient uptake of plants. In this study, the rootstocks significantly affected the root morphology, which is supported by the findings of other studies [38,39]. Liu et al. [40] demonstrated that apple scions grafted onto *Malus prunifolia* or Chistock #1 produced significantly longer, more vigorous root systems than their non-grafted counterparts. Under the low P level condition, the rootstock CL3 performed better than the rootstock CL4. Although the latter showed a little higher evidence of changes in root morphology, there was no significant differences between the two rootstocks under the P-sufficient condition. Furthermore, the rootstock CL3 seemed to also have greater P accumulation and P content in the roots. Leaf functional traits allow for plant survival and maintain plants' ecosystem function [41]. Through grafting, the rootstock has the capability to modify the phenotype of certain scions. The leaf area was affected positively by the rootstock in mango [42]. Specifically, soybean rootstocks can influence the chlorophyll levels in the leaves, thereby impacting the process of leaf aging [43]. Consequently, rootstocks that align with desired traits are often chosen for grafting in agricultural production. For chili peppers, employing drought-resistant rootstocks can shield the plants from oxidative stress due to water scarcity, enhancing their drought tolerance [44]. Different apple rootstock types uniquely influence pollen germination, with specific varieties enhancing cold stress tolerance during flowering [45]. In this study, it was found that the rootstock did not have a significant impact on leaf morphology. However, the rootstocks did have a significant effect on P accumulation in the shoot. This could be attributed to the interaction between the rootstocks and the scions. The influence of the rootstock–scion interaction has been observed and discussed in various crops and fruit trees, including soybeans [30], cotton [31], and apple [34,35,40]. Some researchers have proposed that the apparent interaction between rootstocks and scions for certain traits in the grafted plants is primarily controlled by the rootstocks. For example, potato grafting using late blight-resistant varieties as rootstocks could enhance resistance to late blight in scions derived from susceptible varieties by up-regulating the expression of disease-resistant genes in the scions [41]. The contents of sugars, acids, and volatiles in heterografted tomato pericarps are influenced by the rootstocks [46]. It was further demonstrated that rootstocks play an essential role in grafting. Our results further demonstrated that grafting the scion CL4 to the rootstock CL3 improved the plant growth and increased the P content more than grafting to the rootstocks CL4 and CL53 under low-P stress, which suggests that rootstock may have a crucial regulating effect on grafted *C. oleifera*.

The mechanisms underlying the response to low-P conditions in *C. oleifera* have been widely discussed by many researchers. Su et al. [16] suggested that *CoWRKYs* in the P-efficient CL40 variety of *C. oleifera* may play a crucial role in P deficiency tolerance, being involved in the transportation and recycling of P in the leaves and affecting diverse metabolic pathways. Three hub target genes, *ARF22*, *WRKY53*, and *SCL6*, which may play key roles in CL166 of *C. oleifera* in controlling transcriptomic regulation in response to low Pi stress, were identified [47]. In this study, some pathways were identified related to low P levels in oil-tea species, which were overlooked by previous studies. For example, mitogen-activated protein kinase (MAPK) signaling pathways play a very important role in plant abiotic stress [48]. Under low-P stress, *MAPK* was strongly up-regulated. It significantly modified plant growth under stress conditions and dramatically modified the root architecture through transcriptional regulation of the auxin transport-associated genes [49]. Overexpression of a member of the *MAPK* gene family enhanced tobacco's dry mass production, Pi uptake capacity, and root system architecture establishment under low-Pi stress [50]. It is important to note that our study did not find differences in the expression of transcription factors common in the low-P response. For example, Su et al. [16] indicated that 32 *CoWRKYs* participated in the regulation of responses to P deficiency stress in *C. oleifera*. The complexity of the rootstocks–scion interaction-dominated P response mechanism may account for this phenomenon.

Moreover, this study delivers an integrated, molecular-level roadmap illustrating how distinct rootstocks steer the low-phosphorus adaptive strategies of grafted *C. oleifera*. GO and KEGG profiling exposed sharp functional partitioning between the two graft combinations, with each rootstock–scion pair deploying a unique transcriptional program to confront P limitation. CL4/CL3 preferentially up-regulated phenylpropanoid–flavonoid metabolism, organic-acid biosynthesis, glutathione-mediated antioxidant systems, and phytohormone signaling, thereby constructing a metabolic network centered on efficient P acquisition, mobilization, and translocation. In contrast, CL4/CL4 prioritized defense responses, cell wall remodeling, and *MAPK* cascades, reflecting a 'defend-and-conserve' strategy that enhanced P use efficiency rather than P uptake. This functional separation aligned with phenotypic data: CL4/CL3 exhibited marked increases in root length and surface area and a 28% rise in whole-plant P accumulation under low P, whereas CL4/CL4 improved the PUE while maintaining the biomass. The *MAPK* signaling cascade was the most strikingly differentially expressed pathway. qRT-PCR confirmed that *MAPK* expression in the CL4/CL3 roots reached 2.73-fold that of the control—significantly higher than the 1.28-fold increase in CL4/CL4—indicating that *MAPK*-mediated auxin transport and root architectural remodeling underpin the superior P acquisition of CL4/CL3. Similarly, the stronger up-regulation of the hormone-related gene *PHST* and the defense gene PPI in CL4/CL3 not only intensified the immediate root response to P limitation but may also have activated systemic acquired resistance, indirectly reducing P diverted to defense. Moreover, the up-regulation of the lipid metabolism gene *ELM* and the down-regulation of *FBs* jointly suggest coordinated membrane lipid turnover and P recycling. Collectively, rootstock–scion interactions do not merely sum additive effects; instead, they reprogram the root transcriptional networks to confer either a 'CL3-type P-capture' or a 'CL4-type P-saving' strategy under low P.

5. Conclusions

In this study CL4 scions were grafted onto three rootstocks and exposed to low P or control conditions. CL3 markedly improved scion P uptake and P use efficiency (PUE); under low P it stimulated its own root growth, which translated into greater biomass and total P in the shoot. Transcript profiling further suggested that CL3 primed the scion by up-regulating MAPK, hormone, and defense-related genes, although these candidates still await direct validation. Together the data underline the pivotal role of rootstock choice in grafted crops grown on P-deficient soils and provide a molecular framework for future rootstock–scion studies. The next steps are to confirm that the CL3 low-P advantage persists in the field and to test its performance under drought and in combinations with other commercial scion varieties.

Author Contributions: Z.R., J.Z., L.C. and H.L. conducted the research work and collected the data. Z.R. and J.L. analyzed the data and prepared the manuscript. Y.Z., Q.C., H.W. and W.S. revised the manuscript. D.H. designed and supervised the work. All authors have read and agreed to the published version of the manuscript.

Funding: This research was funded by the National Natural Science Foundation of China (32060362), Ganpo Elite Support Program (20243BCE51086), and The Jiangxi Forestry Bureau Camellia oleifera Research Special Project [YCYJZX(2023)114].

Institutional Review Board Statement: Not applicable.

Informed Consent Statement: Not applicable.

Data Availability Statement: The transcriptomic data presented in this study are deposited in NCBI, accession number PRJNA1328430.

Conflicts of Interest: The authors declare no conflicts of interest.

References

1. Ren, P.; Meng, Y.; Li, B.; Ma, X.; Si, E.; Lai, Y.; Wang, J.; Yao, L.; Yang, K.; Shang, X.; et al. Molecular mechanisms of acclimatization to phosphorus starvation and recovery underlying full-length transcriptome profiling in barley (*Hordeum vulgare* L.). *Front. Plant Sci.* **2018**, *9*, 500. [CrossRef]
2. Navarro, M.; Munné-Bosch, S. Reduced phosphate availability improves tomato quality through hormonal modulation in developing fruits. *J. Plant Growth Regul.* **2022**, *41*, 153–162. [CrossRef]
3. Wang, Y.; Wang, F.; Lu, H.; Liu, Y.; Mao, C. Phosphate uptake and transport in plants: An elaborate regulatory system. *Plant Cell Physiol.* **2021**, *62*, 564–572. [CrossRef] [PubMed]
4. Wang, Z.; Kuo, H.-F.; Chiou, T.J. Intracellular phosphate sensing and regulation of phosphate transport systems in plants. *Plant Physiol.* **2021**, *187*, 2043–2055. [CrossRef]
5. Kuchenbuch, R.O.; Buczko, U. Re-visiting potassium- and phosphate-fertilizer responses in field experiments and soil-test interpretations by means of data mining. *J. Plant Nutr. Soil Sci.* **2011**, *174*, 171–185. [CrossRef]
6. Andersson, H.; Bergström, L.; Djodjic, F.; Ulén, B.; Kirchmann, H. Topsoil and subsoil properties influence phosphorus leaching from four agricultural soils. *J. Environ. Qual.* **2013**, *42*, 455–463. [CrossRef]
7. Cordell, D.; Drangert, J.O.; White, S. The story of phosphorus: Global food security and food for thought. *Glob. Environ. Change* **2009**, *19*, 292–305. [CrossRef]
8. Lott, J.N.A.; Kolasa, J.; Batten, G.D.; Campbell, L.C. The critical role of phosphorus in world production of cereal grains and legume seeds. *Food. Secur.* **2011**, *3*, 451–462. [CrossRef]
9. Vance, C.P.; Uhde-Stone, C.; Allan, D.L. Phosphorus acquisition and use: Critical adaptations by plants for securing a nonrenewable resource. *New Phytol.* **2003**, *157*, 423–447. [CrossRef]
10. Lambers, H.; Finnegan, P.M.; Laliberté, E.; Pearse, S.J.; Ryan, M.H.; Shane, M.W.; Veneklaas, E.J. Phosphorus nutrition of proteaceae in severely phosphorus-impoverished soils: Are there lessons to be learned for future crops? *Plant Physiol.* **2011**, *156*, 1058–1066. [CrossRef] [PubMed]
11. Lynch, J.P.; Brown, K.M. Topsoil foraging—An architectural adaptation of plants to low phosphorus availability. *Plant Soil* **2001**, *237*, 225–237. [CrossRef]
12. Gahoonia, T.S.; Nielsen, N.E. Barley genotypes with long root hairs sustain high grain yields in low-P field. *Plant Soil* **2004**, *262*, 55–62. [CrossRef]
13. Haling, R.E.; Brown, L.K.; Bengough, A.G.; Young, I.M.; Hallett, P.D.; White, P.J.; George, T.S. Root hairs improve root penetration, root–soil contact, and phosphorus acquisition in soils of different strength. *J. Exp. Bot.* **2013**, *64*, 3711–3721. [CrossRef]
14. Zhang, Z.; Zhu, L.; Li, D.; Wang, N.; Sun, H.; Zhang, Y.; Zhang, K.; Li, A.; Bai, Z.; Li, C.; et al. In situ root phenotypes of cotton seedlings under phosphorus stress revealed through RhizoPot. *Front. Plant Sci.* **2021**, *12*, 716691. [CrossRef]
15. Zhang, Z.; Liao, H.; Lucas, W.J. Molecular mechanisms underlying phosphate sensing, signaling, and adaptation in plants. *J. Integr. Plant Biol.* **2014**, *56*, 192–220. [CrossRef]
16. Su, W.; Zhou, Z.; Zeng, J.; Cao, R.; Zhang, Y.; Hu, D.; Liu, J. Genome-wide identification of the WRKY gene family in *Camellia oleifera* and expression analysis under phosphorus deficiency. *Front. Plant Sci.* **2023**, *14*, 1082496. [CrossRef]

17. Jaganath, B.; Subramanyam, K.; Mayavan, S.; Karthik, S.; Elayaraja, D.; Udayakumar, R.; Manickavasagam, M.; Ganapathi, A. An efficient in planta transformation of *Jatropha curcas* (L.) and multiplication of transformed plants through in vivo grafting. *Protoplasma* **2014**, *251*, 591–601. [CrossRef]
18. Warschefsky, E.J.; Klein, L.L.; Frank, M.H.; Chitwood, D.H.; Londo, J.P.; Von Wettberg, E.J.B.; Miller, A.J. Rootstocks: Diversity, domestication, and impacts on shoot phenotypes. *Trends Plant Sci.* **2016**, *21*, 418–437. [CrossRef] [PubMed]
19. Adam, M.B.; Ulas, A. Vigorous rootstocks improve nitrogen efficiency of tomato by inducing morphological, physiological and biochemical responses. *Gesunde Pflanz.* **2023**, *75*, 565–575. [CrossRef]
20. Rodríguez-Gamir, J.; Primo-Millo, E.; Forner, B.J.; Forner-Giner, M.A. Citrus rootstock responses to water stress. *Sci. Hortic.* **2010**, *126*, 95–102. [CrossRef]
21. Feng, J.L.; Yang, Z.J.; Chen, S.P.; El-Kassaby, Y.A.; Chen, H. Signaling pathway in development of *Camellia oleifera* nurse seedling grafting union. *Trees* **2017**, *31*, 1543–1558. [CrossRef]
22. Zeng, J.; Liu, J.; Lian, L.; Xu, A.; Guo, X.; Zhang, L.; Zhang, W.; Hu, D. Effects of scion variety on the phosphorus efficiency of grafted *Camellia oleifera* seedlings. *Forests* **2022**, *13*, 203. [CrossRef]
23. Zeng, J.; Zhao, L.; Liu, J.; Duan, Y.K.; Wang, S.Y.; Wang, Z.L.; Gai, T.T.; Ren, Z.H.; Guo, X.M.; Hu, D.N. Screen of *Camellia oleifera* rootstock genotypes tolerant to lo phosphorus and identification of index tolerant to phosphorus deficiency. *For. Res.* **2019**, *34*, 166–173. [CrossRef]
24. Chen, C.; Chu, Y.; Huang, Q.; Ding, C.; Zhang, W.; Li, B.; Zhang, J.; Su, X. Morphological and physiological plasticity response to low nitrogen stress in black cottonwood (*Populus deltoides* Marsh.). *J. For. Res.* **2022**, *33*, 51–62. [CrossRef]
25. Chai, X.; Wang, X.; Li, H.; Xu, X.; Wu, T.; Zhang, X.; Wang, Y.; Han, Z. Apple scion cultivars regulate the rhizosphere microbiota of scion/rootstock combinations. *Appl. Soil Ecol.* **2022**, *170*, 104305. [CrossRef]
26. Hou, S.; Zhu, Y.; Wu, X.; Xin, Y.; Guo, J.; Wu, F.; Yu, H.; Sun, Z.; Xu, C. Scion-to-rootstock mobile transcription factor *CmHY5* positively modulates the nitrate uptake capacity of melon scion grafted on squash rootstock. *Int. J. Mol. Sci.* **2023**, *24*, 162. [CrossRef]
27. Xie, B.; An, X.H.; Chen, Y.H.; Cheng, C.G.; Kang, G.D.; Zhou, J.T.; Zhao, D.Y.; Li, Z.; Zhang, Y.Z.; Yang, A. Response and adaptability evaluation of different apple rootstocks to continuous phosphorus deficiency. *Sci. Agric. Sin.* **2022**, *55*, 2598–2612. [CrossRef]
28. Zhang, Y.; Liu, X.; Guo, J.; Zhao, J.; Wang, S.; Zheng, Z.; Jiang, Q.; Ren, F. Reponse of root endophytes to phosphorus avaibility in peach rootstocks with constrasting phosphorus-use efficiencies. *Front. Plant Sci.* **2021**, *12*, 2021. [CrossRef]
29. Ibacache, G.A.; Sierra, B.C. Influence of rootstocks on nitrogen, phosphorus and potassium content in petioles of four table grape varieties. *Chil. J. Agric. Res.* **2009**, *69*, 503–508. [CrossRef]
30. Chu, S.; Li, H.; Zhang, X.; Yu, K.; Chao, M.; Han, S.; Zhang, D. Physiological and proteomics analyses reveal low-phosphorus stress affected the regulation of photosynthesis in soybean. *Int. J. Mol. Sci.* **2018**, *19*, 1688. [CrossRef]
31. Kayoumu, M.; Iqbal, A.; Muhammad, N.; Li, X.; Li, L.; Wang, X.; Gui, H.; Qi, Q.; Ruan, S.; Guo, R.; et al. Phosphorus availability affects the photosynthesis and antioxidant system of contrasting low-P-tolerant cotton genotypes. *Antioxidants* **2023**, *12*, 466. [CrossRef] [PubMed]
32. Wu, F.; Li, J.; Chen, Y.; Zhang, L.; Zhang, Y.; Wang, S.; Shi, X.; Li, L.; Liang, J. Effects of phosphate solubilizing bacteria on the growth, photosynthesis, and nutrient uptake of *Camellia oleifera* Abel. *Forests* **2019**, *10*, 348. [CrossRef]
33. Neocleous, D.; Savvas, D. The effects of phosphorus supply limitation on photosynthesis, biomass production, nutritional quality, and mineral nutrition in lettuce grown in a recirculating nutrient solution. *Sci. Hortic.* **2019**, *252*, 379–387. [CrossRef]
34. Hayat, F.; Asghar, S.; Yanmin, Z.; Xue, T.; Nawaz, M.A.; Xu, X.F.; Wang, Y.; Wu, T.; Zhang, X.Z.; Qiu, C.P.; et al. Rootstock induced vigour is associated with physiological, biochemical and molecular changes in "Red Fuji" apple. *Int. J. Agric. Biol.* **2020**, *24*, 1823–1834.
35. Zhou, Y.; Underhill, S.J.R. Expression of gibberellin metabolism genes and signalling component in dwarf phenotype pf breadfruit (*Artocarpus alilis*) plants growing on marang (*Artocarpus odoratissimus*) rootstocks. *Plants* **2020**, *9*, 634. [CrossRef]
36. Tombesi, S.; Johnson, R.S.; Day, K.R.; DeJong, T.M. Interactions between rootstock, inter-stem and scion xylem vessel characteristics of peach trees growing on rootstocks with contrasting size-controlling characteristics. *AoB Plants* **2010**, *2010*, plq013. [CrossRef]
37. Chen, B.; Wang, C.; Tian, Y.; Chu, Q.; Hu, C. Anatomical characteristics of young stems and mature leaves of dwarf pear. *Sci. Hortic.* **2015**, *186*, 172–179. [CrossRef]
38. Adams, S.; Lordan, J.; Fazio, G.; Bugbee, B.; Francescatto, P.; Robinson, T.L.; Black, B. Effect of scion and graft type on transpiration, hydraulic resistance and xylem hormone profile of apples grafted on Geneva®41 and M.9-NIC™29 rootstocks. *Sci. Hortic.* **2018**, *227*, 213–222. [CrossRef]
39. Cichy, K.A.; Snapp, S.S.; Kirk, W.W. Fusarium root rot incidence and root system architecture in grafted common bean lines. *Plant Soil* **2007**, *300*, 233–244. [CrossRef]

40. Liu, X.; Zhang, M.; Hu, H.; Li, Z.; Xu, X.; Wang, Y.; Han, Z. Drought tolerance and impacts of four rootstock genotypes on the morphology, yield and fruit quality of Fuji scion apple under drought conditions. *Hortic. Environ. Biotechnol.* **2024**, *65*, 491–500. [CrossRef]
41. Li, Y.; Zhao, D. Transcriptome analysis of scions grafted to potato rootstock for improving late blight resistance. *BMC Plant Biol.* **2021**, *21*, 272. [CrossRef]
42. Rebolledo-Martínez, A.; Peralta-Antonio, N.; Rebolledo-Martínez, L.; Becerril-Román, E.A.; Rebolledo-García, R.L. Effect of rootstock in tree growth, dry matter, flowering, yield and quality of 'Manila' mango. *Sci. Hortic.* **2019**, *251*, 155–161. [CrossRef]
43. Ookawa, T.; Tomita, N.; Hirasawa, T. Interaction of scion and stock on leaf senescence of soybean plants grafted at mid-stem during ripening. *Plant Prod. Sci.* **2005**, *8*, 32–37. [CrossRef]
44. López-Serrano, L.; Canet-Sanchis, G.; Vuletin Selak, G.; Penella, C.; San Bautista, A.; López-Galarza, S.; Calatayud, Á. Pepper rootstock and scion physiological responses under drought stress. *Front. Plant Sci.* **2019**, *10*, 38. [CrossRef]
45. Zebro, M.; Heo, J.Y. Rootstock type differentially affects pollen germination characteristics and chilling stress response at flowering time in 'Fuji' spple. *Appl. Fruit Sci.* **2023**, *66*, 333–340. [CrossRef]
46. Zhou, Z.; Yuan, Y.; Wang, K.; Wang, H.; Huang, J.; Yu, H.; Cui, X. Rootstock-scion interactions affect fruit flavor in grafted tomato. *Hortic. Plant J.* **2022**, *8*, 499–510. [CrossRef]
47. Chen, J.; Han, X.; Ye, S.; Liu, L.; Yang, B.; Cao, Y.; Zhuo, R.; Yao, X. Integration of small RNA, degradome, and transcriptome sequencing data illustrates the mechanism of low phosphorus adaptation in *Camellia oleifera*. *Front. Plant Sci.* **2022**, *13*, 932926. [CrossRef]
48. Moustafa, K.; AbuQamar, S.; Jarrar, M.; Al-Rajab, A.J.; Trémouillaux-Guiller, J. MAPK cascades and major abiotic stresses. *Plant Cell Rep.* **2014**, *33*, 1217–1225. [CrossRef]
49. Hao, L.; Wen, Y.; Zhao, Y.; Lu, W.; Xiao, K. Wheat mitogen-activated protein kinase gene *TaMPK4* improves plant tolerance to multiple stresses through modifying root growth, ROS metabolism, and nutrient acquisitions. *Plant Cell Rep.* **2015**, *34*, 2081–2097. [CrossRef] [PubMed]
50. Bai, J.; Xie, Y.; Shi, M.; Yao, S.; Lu, W.; Xiao, K. *TaMPK2B*, a member of the MAPK family in T. aestivum, enhances plant low-Pi stress tolerance through modulating physiological processes associated with phosphorus starvation defensiveness. *Plant Sci.* **2022**, *323*, 111375. [CrossRef] [PubMed]

life

MDPI

Review

Spent Mushroom Substrate-Derived Biochar and Its Applications in Modern Agricultural Systems: An Extensive Overview

Worawoot Aiduang [1,2], Kritsana Jatuwong [1,2], Tanongkiat Kiatsiriroat [3], Wassana Kamopas [4], Pimsiri Tiyayon [5], Rotsukon Jawana [6], Orlavanh Xayyavong [2] and Saisamorn Lumyong [2,7,8,*]

1 Office of Research Administration, Chiang Mai University, Chiang Mai 50200, Thailand; worawoot.aiduang@cmu.ac.th (W.A.); kritsana.ja@cmu.ac.th (K.J.)
2 Department of Biology, Faculty of Science, Chiang Mai University, Chiang Mai 50200, Thailand; xorlavanh@yahoo.com
3 Department of Mechanical Engineering, Faculty of Engineering, Chiang Mai University, Chiang Mai 50200, Thailand; tanongkiat_k@yahoo.com
4 Multidisciplinary Research Institute, Chiang Mai University, Chiang Mai 50200, Thailand; wassana.kamopas@cmu.ac.th
5 School of Agricultural Resources, Chulalongkorn University, Bangkok 10330, Thailand; pimsiri.t@chula.ac.th
6 Energy Research and Development Institute-Nakornping, Chiang Mai University, Chiang Mai 50200, Thailand; rotsukon.j@cmu.ac.th
7 Center of Excellence in Microbial Diversity and Sustainable Utilization, Chiang Mai University, Chiang Mai 50200, Thailand
8 Academy of Science, The Royal Society of Thailand, Bangkok 10300, Thailand
* Correspondence: saisamorn.l@cmu.ac.th; Tel.: +66-956-914-928

Abstract: Spent mushroom substrate (SMS), a nutrient-dense byproduct of mushroom cultivation, has emerged as a promising feedstock for biochar production, offering a sustainable solution to modern agricultural and environmental challenges. This review explores SMS properties, its conversion into biochar, and its various applications. Due to its lignocellulosic structure, high organic matter (OM), and essential nutrients, SMS is ideal for pyrolysis, a process that enhances biochar's porosity, nutrient retention, and carbon stability. These properties improve soil fertility, water retention, microbial activity, and plant growth while also contributing to climate change mitigation through carbon sequestration. SMS-derived biochar stands out for its superior benefits, including a balanced pH, a rich nutrient profile, and the ability to adsorb heavy metals, which mitigates soil and water contamination and minimizes toxic risks in the food chain. By enhancing soil structure, nutrient cycling, and moisture retention, SMS-derived biochar supports sustainable farming practices that reduce chemical fertilizer use and boost climate resilience. Beyond soil applications, SMS-derived biochar is effective in wastewater treatment, mitigating plant diseases, and improving mushroom cultivation substrates, thereby enhancing mycelial growth and productivity. Economically, it is a cost-effective alternative due to the abundant availability and inexpensive nature of SMS. Nevertheless, challenges still exist, particularly in optimizing production methods and ensuring consistency in biochar properties, influenced by variations in pyrolysis conditions and SMS types. Advances in production technology and sustainable practices are vital for scaling up SMS-derived biochar production. This paper emphasizes the transformative potential of SMS-derived biochar, advocating for its integration into circular economy frameworks and sustainable agricultural systems. Recommendations for future research and policy support are provided to maximize the ecological and economic benefits of SMS-derived biochar, fostering its widespread adoption in global agricultural and environmental strategies.

https://doi.org/10.3390/life15020317

Keywords: SMS biochar potential; carbon sequestration; soil amendment; sustainable agriculture; sustainable development goals (SDGs) 2 and 15

1. Introduction

SMS, a byproduct of mushroom cultivation, represents both a valuable resource and a waste management challenge in modern agricultural systems [1]. With an organic-rich composition and a nutrient profile often enriched with nitrogen (N), phosphorus (P), potassium (K) as well as other minerals (calcium (Ca), magnesium (Mg), sulfur (S), iron (Fe), manganese (Mn), zinc (Zn), copper (Cu), and sodium (Na) [1,2], SMS has garnered interest for its potential as a raw material in biochar production. However, given the extensive global production of mushrooms (estimated at approximately 48.34 million metric tons), particularly in the top 10 mushroom-producing countries in the world, including China, Italy, the United States of America, the Netherlands, Poland, Spain, France, Iran, Canada, and the United Kingdom, SMS disposal presents considerable environmental challenges [3]. As a significant byproduct of mushroom farming, SMS is frequently considered waste. If improperly managed, it can contribute to greenhouse gas emissions, leach nutrients into groundwater, and accumulate as waste, resulting in environmental hazards through the development of pathogenic microflora and the spread of fungal diseases. It also increases eutrophication and adds to the financial, logistical, and ecological burdens for mushroom farms [4,5]. Crucially, unsuitable disposal can result in disagreeable smells and draw insects, causing additional issues for the environment and community wellness [6]. Despite these challenges, SMS holds considerable potential as a valuable feedstock and carbon source, particularly for biochar production, which offers a sustainable solution to waste disposal while providing numerous agricultural benefits. Converting SMS into biochar not only addresses these environmental concerns but also yields a material with wide-ranging agricultural benefits, aligning with the principles of sustainable and circular economy practices [7–9].

Biochar, a form of carbon-rich material produced through pyrolysis, has shown immense promise for soil health, pollutant remediation, and water management, improves material sustainability, and even supports urban agriculture and green infrastructure [10]. When applied to agricultural soils, SMS-derived biochar can improve soil fertility, water retention, and carbon sequestration, making it a valuable asset in sustainable farming practices and modern agricultural systems [7]. The pyrolysis process, which heats organic material in low-oxygen conditions, allows for SMS to be transformed into a stable carbon form that is resistant to decomposition, locking carbon into the soil for extended periods [10,11]. Optimal pyrolysis conditions for SMS-derived biochar production involve carefully controlled temperatures and residence times, which affect the physicochemical properties of the resulting biochar [11,12]. This thermochemical conversion process requires energy and technical considerations; yet, it has proven to be economically viable for agricultural use, particularly when offset by the benefits to soil and crop productivity [7,13,14].

The physicochemical properties of SMS-derived biochar, including high porosity and surface area, pore size distribution, high carbon content, as well as nutrient and moisture retention capabilities, make it particularly effective in various agricultural applications [13,14]. Its chemical characteristics, such as pH, cation exchange capacity (CEC), and the potential for heavy metal adsorption, distinguish it from biochars derived from other organic wastes, positioning it as a robust soil amendment for multiple agricultural purposes [13–15]. SMS-derived biochar's stability and persistence in the soil enable it to improve soil structure and nutrient availability, thus enhancing plant growth, yield, and biochemical response, while simultaneously resulting in reduced incidence of disease [7,16,17]. Additionally, the biochar's ability to adsorb heavy metals and organic pollutants opens opportunities for its use in agricultural wastewater treatment, protecting water resources from contamination by pesticides [16].

Currently, SMS-derived biochar's applications extend beyond soil health and include carbon sequestration, nutrient retention, improved plant growth and biochemical response, and even an enhancement of mushroom production systems [1,7,14,16,18]. Moreover, several previous studies have indicated that several biochars offer a natural approach to plant disease management [17,19]. This presents a good guideline for further utilization of SMS-derived biochar in the future. Several comparative studies reveal that SMS-derived biochar performs similarly or even superiorly to traditional biochars made from wood or agricultural residues, largely due to SMS's unique nutrient profile and structural properties [14,20]. This performance, coupled with its abundance and low cost, supports SMS-derived biochar's environmental and economic viability as an amendment in modern farming systems [14,15,21].

Despite its potential, challenges remain in standardizing SMS-derived biochar production and addressing environmental concerns [22]. The influence of biochar on soil microbial communities, potential contaminants, and regulatory needs are areas that require further exploration to maximize the safety and efficacy of SMS-derived biochar in agricultural systems. Moreover, long-term studies are essential to understand its persistence and impact on soil ecology over time. Therefore, this review comprehensively examines SMS-derived biochar's properties, production methods, and diverse applications in modern agriculture, exploring its potential benefits, limitations, and future directions. We also investigate novel opportunities in precision agriculture, water management, and pollution control, positioning SMS-derived biochar as a versatile and impactful component in sustainable agricultural practices.

2. Properties of Spent Mushroom Substrate

2.1. Overview of the Composition and Nutritional Profile of SMS

SMS is a nutrient-rich organic byproduct generated after harvesting edible and medicinal mushrooms, including widely cultivated genera such as *Agaricus*, *Auricularia*, *Coprinus*, *Flammulina*, *Ganoderma*, *Hericium*, *Lentinula*, *Lentinus*, *Pleurotus*, and *Tremella* [7,12,13,16,20,23–26]. This byproduct is composed of residual mycelium, lignocellulosic compounds (lignin: 25–39.8%, hemicellulose: 19–34%, and cellulose: 38–48.7%), bioactive compounds, fibers, proteins, non-cell-wall carbohydrates, and essential minerals derived from agricultural residues such as sawdust, corn cob, cottonseed hull, straw, and wood chips, along with amendments like gypsum, lime, and peat [1,6,27–29].

Each SMS batch offers unique physicochemical properties, enriched with both macronutrients and micronutrients essential for plant growth (Table 1). For example, SMS holds high OM (58.97 to 92.73%) and vital nutrients like N (0.46–2.72%), P (0.02–0.84%), and K (0.02–2.40%). It is also a rich source of micronutrients such as Ca (0.52–10.92%), Mg (0.09–0.91%), S (0.2–1.6%), Fe (7.60–920 $mg \cdot kg^{-1}$), Mn (100–1600 $mg \cdot kg^{-1}$), Zn (26–220 $mg \cdot kg^{-1}$), Cu (10.80–79.67 $mg \cdot kg^{-1}$), and Na (433.33–6693.33 $mg \cdot kg^{-1}$), further enhancing its effectiveness as a soil amendment.

Table 1. The physicochemical properties of SMS remains from each mushroom production.

Properties	Spent Mushroom Substrates							
	Agaricus bisporus	*Coprinus comatus*	*Flammulina velutipes*	*Ganoderma lucidium*	*Hericium erinaceus*	*Lentinula edodes*	*Pleurotus ostreatus*	*Tremella fuciformis*
EC (dS m^{-1})	4.83–6.60	0.88	-	1.92	1.08	1.96	0.89–4.01	-
pH	5.02–8.05	5.80	5.83	7.10	4.24	3.79–4.33	5.54–6.49	6.12
OM (%)	66.98	-	-	58.97	90.09	92.73	61.17–91.87	-
OC (%)	23.40–54.30	-	41.24	34.20–44.82	52.25	38.96–53.78	32.61–53.29	40
C/N ratio	11.50–25.43	-	29.88	27.33–74.35	92.60	26.69–40.95	19.00–79.31	19.51
Total N (%)	0.92–2.72	-	1.38	0.46–1.64	0.57	1.12–1.46	0.56–1.72	2.05
Total P (%)	0.21–0.84	-	0.04	0.09–0.36	0.10	0.02–0.07	0.05–0.77	0.03
Total K (%)	0.43–2.40	0.85	0.02	0.48–1.03	0.35	0.10–0.21	0.23–1.90	0.04
Ca (%)	4.70–10.80	5.15	-	2.66–10.92	0.84	0.52	0.68–10.44	-
Mg (%)	0.60–0.91	0.30	-	0.24–0.28	0.18	0.45	0.09–0.51	-
S (%)	-	-	0.7	-	-	0.91	0.20	1.60
Fe (mg·kg^{-1})	174.27–920	-	-	38.27–60	27.93	21.40	7.60–30.80	-
Mn (mg·kg^{-1})	241.20–1600	-	-	100–291.80	282.80	225.40	134.87–225.93	-
Zn (mg·kg^{-1})	136.41–220	-	-	37.47	26.00	56.47	28.20–49.20	-
Cu (mg·kg^{-1})	50.70–79.67	-	-	14.93	11.60	25	10.80–11.60	-
Na (mg·kg^{-1})	600–6693.33	1900	-	886.67	660	1100	433.33–1233.33	-
WHC (%)	277.5	-	-	231.39	324.10	366.30	347.31–699.68	-
References	[16,23–25]	[13]	[7]	[24,30]	[24]	[7,12,24]	[20,24,26]	[7]

Note: EC refers to electrical conductivity; C/N represents carbon-to-nitrogen ratio; and WHC refers to water-holding capacity.

Notably, SMS has excellent physical properties, particularly its exceptional WHC, which ranges from 231.39 to 699.68%, significantly higher than traditional growing media like peat (~209.96%) or mixed plant growth media (~75.23%) [24]. This superior water retention capacity is invaluable for agricultural applications. Moreover, the residual fungal components in SMS contribute to a rich microbial profile, promoting soil biodiversity [31]. Its organic carbon (OC) and C/N ratio, ranging from 23.40 to 54.30% and 11.50:1 to 92.60:1, respectively, depending on the original substrate, makes it ideal for composting and biochar production. Similarly, its pH range (3.79–8.05) and EC (0.88–6.60 dS m^{-1}) demonstrate its versatility across different soil types and crops. As a sustainable and non-toxic resource, SMS aligns with organic farming practices and supports the global movement toward sustainable agriculture, helping to foster more resilient and productive ecosystems [6].

2.2. Benefits of Biochar Conversion

SMS offers several advantages as a biochar feedstock, making it a valuable resource for sustainable biochar production [32]. As a material rich in lignocellulose, SMS has potential carbon sources that enable efficient carbonization during the pyrolysis process, resulting in its stable high-quality biochar [6,33]. This carbon-rich composition makes SMS-derived biochar an effective medium for carbon sequestration in soils or other growing substrates and could potentially mitigate climate change [34].

In addition, SMS contains a range of essential primary and secondary macronutrients, which are largely preserved during conversion to biochar. This nutrient retention enhances its role as a soil amendment by improving soil fertility and promoting plant growth when applied to agricultural systems [20,35,36]. In mushroom cultivation, studies also indicate that incorporating SMS-derived biochar into growing substrates can enhance mycelial growth and overall yield, demonstrating its utility across agricultural practices [14, 20,36]. The physical properties of SMS, such as its natural porosity and high surface area, are maintained when it is converted to biochar, providing an ideal structure for water retention and fostering beneficial microbial habitats in soils [14,36]. Moreover, producing biochar from SMS addresses the environmental challenges of SMS disposal, transforming agricultural waste into a valuable byproduct and adding significant value to mushroom cultivation residues [6,27]. These combined benefits highlight SMS as an effective and sustainable feedstock for biochar, supporting both agricultural productivity and environmental health.

2.3. Environmental Concerns and Potential Impacts of SMS Disposal on the Ecosystem

The disposal of SMS presents several environmental challenges that can lead to significant ecological impacts [6]. As a major byproduct of mushroom cultivation, SMS is often regarded as waste, resulting in large quantities accumulating on farms or in landfills [37]. This accumulation can lead to various issues, such as the leaching of nutrients into water systems, which contributes to eutrophication. This process promotes harmful algal blooms that deplete oxygen levels in water environments, adversely affecting aquatic life and disrupting ecosystems [31,38]. Additionally, improper disposal can lead to unpleasant odors and attract pests, creating further environmental and public health concerns [6].

Additionally, SMS is rich in organic matter, and if not managed properly, it can undergo anaerobic decomposition, releasing potent greenhouse gases like methane (CH_4), nitrous oxide (N_2O), ammonia (NH_3), and nitrate (NO^{3-}), which have a far greater impact on climate change than carbon dioxide [37,39]. Without effective management practices, the environmental footprint of SMS can be considerable, highlighting the urgent need for sustainable waste management solutions [27]. Converting SMS into biochar, compost, or other value-added products offers a way to address these environmental challenges, transforming waste into a sustainable resource that can enhance agricultural practices and promote environmental health [6,27,28].

3. Production of Biochar from Spent Mushroom Substrate

3.1. Pyrolysis Process

Pyrolysis is a thermochemical process that converts biomass into biochar by heating it under controlled oxygen-limited conditions. As pyrolysis occurs at varying temperatures, the surface structure of the obtained biochar can change, influenced by the additional breakdown of lignin and other organic components in the feedstock [40]. The process generally operates within a temperature range of 300 to 1300 °C, where the medium (300–800 °C) and high temperatures (800–1300 °C) are tailored to optimize biochar properties depending on the desired application [41–43]. For SMS-derived biochar, pyrolysis temperatures typically range between 300 and 900 °C [12,14,16,32,44–48], with optimal biochar properties generally recorded at temperatures between 500 and 750 °C [14,44,49–51]. At these temperatures, the carbonization process enhances the stability and quality of the resulting biochar, improving its adsorption capabilities and nutrient retention [49]. The pyrolysis yields three primary products: gases (including carbon dioxide, carbon monoxide, and hydrocarbons), liquids (such as organic solutions and tar), and solids (char) [52]. The specific proportions and compositions of these products vary based on the type of biomass used, the pyrolysis method, and the reactor design employed [53].

Various types of reactors can be utilized for the pyrolysis of SMS, including batch, continuous, and fluidized bed reactors. Batch reactors are straightforward and allow for flexibility in feedstock processing but may have lower efficiency. Continuous reactors, on the other hand, facilitate a more efficient and scalable process by continuously feeding SMS into the reactor, thereby enhancing productivity. Fluidized bed reactors are particularly advantageous as they provide excellent heat transfer and uniform temperature distribution, which are critical for consistent biochar quality [54]. By carefully selecting the pyrolysis conditions and reactor type, producers can optimize the characteristics of SMS-derived biochar for various applications.

3.2. Influence of Parameters on Biochar Properties

The production of SMS-derived biochar is significantly influenced by various pyrolysis parameters, including the conditions used in pyrolysis (temperature, pressure, heating rate, final temperature, residence time, reactor atmosphere, and feed system), the type of reactor (which affects the heating rate and process duration), and the characteristics of the feedstock (particle size, type, and composition) [51,53,55]. Higher pyrolysis temperatures typically enhance the carbon content, surface area, pore size, and some nutrients, like the P and K contents of the resulting biochar, promoting a more stable structure [16,51]. For example, biochar produced at temperatures exceeding 500 °C typically exhibits high hydrophobicity, increased surface area, and higher micropore volume, making it suitable for adsorbing organic pollutants. In contrast, biochar produced at lower temperatures (below 500 °C) retains more oxygenated functional groups, which are advantageous for immobilizing inorganic pollutants [56,57]. These characteristics highlight the importance of optimizing pyrolysis conditions to tailor SMS-derived biochar to specific environmental and agricultural applications.

Residence time during pyrolysis plays a vital role in determining the quality and properties of SMS-derived biochar. Extended residence times can enhance carbonization but may also lead to the loss of volatile compounds, reducing nutrient availability, and potentially degrading overall biochar quality [10,58]. Additionally, variations in the feedstock type, such as different mushroom species or substrates, can affect the physicochemical properties of the biochar, including its pH, CEC, essential nutrients, and overall stability [7,10]. Understanding these mechanisms is essential for optimizing biochar production processes to achieve specific characteristics suited for various applications, from soil enhancement to environmental remediation [59]. Overall, careful control of pyrolysis conditions can produce biochar from SMS with desirable properties, while preparing the raw material by reducing its moisture content to below 15% before firing ensures higher biochar yield and quality, extends furnace lifespan, and minimizes emissions. This enhances efficiency in agricultural and ecological applications as the carbon in biochar also contributes to reducing greenhouse gases [12,60].

Furthermore, the effectiveness of SMS-derived biochar can be enhanced through pre- and post-treatment methods that modify its surface chemistry, porosity, and adsorption capacity for targeted applications [61]. Pre-treatment often involves impregnating biochar feedstocks with metallic (e.g., aluminum, iron, manganese, and zinc) or non-metallic (e.g., nitrogen, sulfur, phosphorous, and boron) elements before pyrolysis, creating engineered carbon structures that boost adsorption efficiency [62,63]. However, concerns over metal leaching have led to interest in co-doping strategies, where metal/non-metal combinations enhance stability [61,64]. Post-treatment methods, such as acid or alkaline washing, alter surface functional groups and increase surface area [65]. Additionally, steam modification and gas purging (CO_2 and NH_3) further optimize micropore structures and chemical properties [66]. These modifications allow for biochar to be tailored to specific contaminants, making it a highly adaptable material for environmental remediation and soil enhancement in agricultural systems.

3.3. Energy Efficiency and Economic Viability

Producing biochar from SMS presents an opportunity for energy efficiency (biochar, bio-oil, and bio-gas) while contributing to the commercial, industrial, and economic benefits of society [9,27,67]. The carbonization process, which involves pyrolyzing SMS at high temperatures, not only transforms this agricultural waste into a valuable resource but also captures the energy released during the process. This captured energy can be utilized to power biochar production itself, thereby improving the overall energy efficiency of the operation [13,68,69]. Studies indicate that SMS-derived biochar has an energy content ranging from 17.32 to 22.72 MJ/kg, highlighting its potential as a renewable energy source [32,44,70,71].

Utilizing inexpensive lignocellulosic residues offers a cost-effective and environmentally friendly approach to substrate production, facilitating their efficient management and valorization [6]. Economically, producing biochar from SMS is notably more affordable ($60–117.50/ton or $0.06–0.1175/kg) compared to conventional feedstocks such as oak wood ($0.77), coconut shell ($0.80), *Oiltea camellia* shell ($0.67), chicken manure ($1.30), and virgin wood feedstock ($17.8) [14,50,72,73]. The use of widely available agricultural byproducts significantly lowers raw material costs, benefiting smallholder farmers and increasing the profitability of mushroom producers [14]. Cost analyses reveal that SMS-derived biochar production can be competitive, with production costs varying based on technology and production scale [14,72,73]. Moreover, the potential for additional revenue streams through the sale of biochar as a soil amendment or as a carbon offset can further enhance its economic viability. The combination of energy recovery, reduced production costs, and the generation of valuable byproducts positions SMS-derived biochar as a sustainable and economically attractive alternative in the agricultural and energy sectors [8,74]. As research continues to optimize environmentally friendly production methods and develop processes that reduce carbon dioxide emissions into the environment, such as utilizing waste heat to pre-dry high-moisture raw materials before pyrolysis, the anticipated economic benefits are expected to increase, making SMS-derived biochar a promising solution for sustainable agricultural practices and waste management.

4. Physicochemical and Biological Properties of SMS-Derived Biochar

4.1. Structural, Physical, and Chemical Properties

The physicochemical properties of SMS-derived biochar are pivotal in determining its effectiveness as both a soil amendment and environmental remediation tool [7]. The characteristics of biochar vary depending on the mushroom species from which the SMS is derived (Table 2), but some common features stand out. One key attribute is its pore structure, which significantly impacts its specific surface area (SSA) and particle size (PS) [7,14]. SMS-derived biochar often exhibits a substantial SSA, ranging from 3.09 to 1095 m^2g^{-1}, enabling it to efficiently adsorb nutrients and contaminants [7,16,20,48]. Its PS, typically between 0.92 and 50 mm, enhances interactions with soil components while promoting root penetration and aeration in plants [16,20,33,46,48].

Chemically, SMS-derived biochar features a pH range of 6.49 to 12.31, making it suitable for neutralizing acidic soils. Its EC, measured at 2.09 to 3.54 dS m^{-1}, indicates efficient ionic mobility in the soil. Nutritionally, it is enriched with macronutrients such as N (0.44–2.71%), P (0.03–3.80%), and K (0.08–1.39%), along with significant micronutrients, including Ca (0.40–9.69%), Mg (0.50–1.70%), and trace elements like Fe (1500–24,000 $mg{\cdot}kg^{-1}$), Mn (10–1200 $mg{\cdot}kg^{-1}$), Zn (84–1000 $mg{\cdot}kg^{-1}$), Cu (10 $mg{\cdot}kg^{-1}$), and Na (290–1100 $mg{\cdot}kg^{-1}$). This nutrient wealth supports plant growth and fosters beneficial microbial activity in the soil.

Another standout property is its CEC, which measures its ability to retain and release nutrients to plants. With CEC values between 19.37 and 32.29 cmol kg^{-1}, SMS-derived biochar serves as an effective nutrient reservoir [7]. Additionally, its high carbon content ensures exceptional stability in the soil, resisting microbial degradation and maintaining its benefits for a long time [7,14]. Interestingly, the stability and high carbon content of SMS-derived biochar make it suitable for reuse in mushroom cultivation. Its strong carbon–carbon double bonds resist reactions with water or lime during the preparation of mushroom substrates at room temperature [14,20,75]. This durability and versatility position SMS-derived biochar as an ideal solution for sustainable agriculture and environmental management, offering long-term benefits for soil health, productivity, and ecosystem resilience.

Table 2. The physicochemical properties of SMS-derived biochar from each mushroom species after the pyrolysis process.

Properties	SMS-Derived Biochar							
	Agaricus bisporus	*Coprinus comatus*	*Flammulina velutipes*	*Ganoderma lucidum*	*Lentinula edodes*	*Pleurotus* sp.	*Pleurotus ostreatus*	*Tremella fuciformis*
EC (dS m^{-1})	2.09	3.54	-	-	-	2.81	-	-
pH	8.10	10.05	9.84–11.64	7.78–11.38	6.49–12.07	9.27	9.58–12.31	11.32–11.63
C (%)	62.06	-	54.85–58.61	53.23–70.72	49.44–58.65	50	37.49–69.40	52.60–54.32
C/N ratio	8.67	-	33.9–47.44	27.30–27.73	21.64–24.47	172.41	53.40–85.20	25.89–29.30
CEC (cmol kg^{-1})	-	-	25.71–32.29	-	19.37–27.15	-	-	26.71–28.98
Total N (%)	-	-	1.16–1.67	1.95–2.55	1.09–2.71	-	0.44–1.30	1.80–2.07
Total P (%)	-	-	0.11–0.12	0.71–0.95	0.03–0.06	0.58	3.80	0.06–0.07
Total K (%)	-	1.19	0.08–0.10	0.84–1.26	0.11–1.25	1.39	-	0.12–0.15
Ca (%)	-	9.69	-	3.75–8.94	7.48–11.35	13.36	0.40	-
Mg (%)	-	0.81	-	0.50–0.72	1.09–1.12	1.70	-	-
S (%)	-	-	1.32–1.60	-	0.04–0.49	1.01	-	0.07–0.14
Fe ($mg \cdot kg^{-1}$)	3500	-	-	1500–2500	-	1895	24000	-
Mn ($mg \cdot kg^{-1}$)	1200	-	-	10–20	-	33	-	-
Zn ($mg \cdot kg^{-1}$)	1000	-	-	-	-	84	-	-
Cu ($mg \cdot kg^{-1}$)	-			-	-	10	-	-
Na ($mg \cdot kg^{-1}$)	-	1110	-	290–660	520–920	500	-	-
SSA (m^2 g^{-1})	5.10	-	148.3–210.57	7.18–105.04	3.09–73.56		17.5–1095	24.47–51.73
PS (nm)	0.92	-	-	2–50	2–50	-	2.75–21.8	-
References	[16]	[13]	[7]	[30,46]	[7,12,33,46]	[36]	[20,33,48]	[7]

4.2. *Capability of Heavy Metal Adsorption*

Biochar derived from SMS has shown significant potential in addressing heavy metal contamination in soils and water sources, offering a sustainable and effective solution for soil remediation and environmental protection [7,16,49]. Several studies have demonstrated that biochar, regardless of its source, can reduce the bioavailability of heavy metals in contaminated soils and polluted wastewater, thus limiting their potential to enter the food chain and reducing ecological risks [7,13,15,16,76]. Specifically, SMS-derived biochar has been recognized for its ability to adsorb and immobilize heavy metals such as cadmium (Cd), chromium (Cr), Cu, lead (Pb), and Zn, making it an excellent option for treating contaminated water sources, as well as for protecting human health from long-term exposure to these toxic elements [15,41,77].

The unique physicochemical properties of SMS-derived biochar, including its high surface area, porous structure, and active functional groups, contribute to its exceptional ability to bind heavy metals [49]. These characteristics enable SMS-derived biochar to reduce the mobility and bioavailability of these metals in the soil, which prevents their uptake by plants and reduces the associated environmental risks [15,41,77]. Furthermore, the stable carbon matrix and high carbon content of SMS-derived biochar enhance its long-term effectiveness by preventing the leaching of metals into groundwater, offering a lasting solution to soil contamination [59,78]. The nutrient-rich composition of SMS-derived biochar also supports the growth of beneficial microorganisms [7], which can further aid in the degradation or transformation of pollutants, leading to a healthier and safer soil environment [31,55].

The adsorption capabilities of biochar produced from SMS for the removal of heavy metals are comprehensively described in Table 3, demonstrating its potential as an economically feasible and environmentally conscious remediation method. The data show a considerable variation in adsorption efficiency depending on the type of mushroom substrate used and the specific heavy metal being targeted.

Table 3. Overview of the SMS-derived biochar's adsorption capabilities for the removal of heavy metals following modifications and summaries from Madzin et al. [15] and Jiang et al. [79].

Sources of SMS-Derived Biochar	Adsorption Capacity of Heavy Metals ($mg \cdot g^{-1}$)					References
	Cadmium	Chromium	Copper	Lead	Zinc	
Spent mushroom substrate from *A. bisporus*	64.8	-	68.1	-	55.2	[47,80]
Spent mushroom substrate from *A. auricula*	114.6	118.0	-	-	-	[79]
Spent mushroom substrate from *L. edodes*	-	-	-	398	-	[33]
Spent mushroom substrate from *P. ostreatus*	55.95	-	-	326	-	[33,77]
Spent mushroom substrate from *G. lucidum*	23.81–75.82	-	-	141.59–262.76	-	[30]
Spent mushroom compost	-	-	52.6–364	564	332	[49,81]

Among the various SMS-derived biochars, *A. bisporus*-derived biochar exhibited strong adsorption capacities for cadmium (64.8 mg g^{-1}), copper (68.1 mg g^{-1}), and zinc (55.2 mg g^{-1}), making it a viable option for treating cadmium- and copper-contaminated environments. Similarly, *A. auricula*-derived biochar demonstrated the highest chromium (118.0 mg g^{-1}) and cadmium (114.6 mg g^{-1}) removal efficiencies, indicating its potential application in chromium-contaminated sites. The biochar produced from *L. edodes* and *P. ostreatus* showed remarkable adsorption capacities for lead, with values reaching 398 mg g^{-1} and 326 mg g^{-1}, respectively. Notably, *G. lucidum*-derived biochar exhibited a wide range of cadmium (23.81–75.82 mg g^{-1}) and lead (141.59–262.76 mg g^{-1}) adsorption efficiencies, suggesting that its adsorption performance may be influenced by processing conditions or biochar modification techniques. In addition, spent mushroom compost-derived biochar demonstrated exceptionally high adsorption capacities across multiple heavy metals, including copper (52.6–364 mg g^{-1}), lead (564 mg g^{-1}), and zinc (332 mg g^{-1}). These results underscore its broad-spectrum potential for wastewater treatment and heavy metal detoxification.

Overall, the findings demonstrate that SMS-derived biochar has an interesting potential as an efficient heavy-metal-removal adsorbent, with some mushroom substrates showing greater attraction for various metal ions. Further research into improving the conditions and modification methods used to produce biochar may increase its adsorption efficiency, providing the possibility for scalable uses in environmental remediation and sustainable crop waste management.

4.3. Long-Term Stability and Environmental Persistence

The stability and environmental persistence of SMS-derived biochar are essential for its effectiveness as a soil amendment [59]. Research studies indicate that SMS-derived biochar is highly stable, owing to its unique physicochemical properties, such as an elevated carbon content and a porous structure. These features enhance its resistance to microbial degradation, enabling it to remain in the soil for prolonged periods while gradually enhancing soil health and fertility [14,55,59]. The slow decomposition of biochar not only aids in carbon sequestration but also minimizes nutrient leaching, thereby improving nutrient availability for plants over time [34]. Additionally, its incorporation into mushroom-growing substrates also boosts yields, retains substrate moisture, and minimizes nutrient loss [14]. One might assume that this helps maintain harvest stability and yield quality consistency. Consequently, integrating SMS-derived biochar into agricultural practices offers a sustainable approach to improving soil quality while contributing to long-term environmental objectives.

4.4. Beneficial Microbial Communities

SMS-derived biochar promotes the growth of beneficial microbial communities in the soil, which in turn supports enhanced plant health and growth. Its porous structure and nutrient-rich composition create an ideal habitat for microorganisms, particularly nitrogen-fixing bacteria such as plant growth-promoting rhizobacteria (PGPR), which play a key role in nutrient treatment and plant health [16]. The porous nature of biochar, consisting of micropores (less than 2 nm), mesopores (2–50 nm), and macropores (more than 50 nm), offers an ideal habitat for a variety of microorganisms, such as bacteria, protozoa, and fungi [82]. These microhabitats support microbes from leaching, competition, predators, toxins, and moisture loss, fostering their activity and interaction [83]. The macropores of biochar are well-suited for microbial colonization, while mesopores and micropores store dissolved nutrients and water, supplying the resources needed for microbial metabolism [84,85]. This synergy contributes to improved plant growth and resilience, aligning with recent advancements in microbe-assisted plant development strategies [86]. Moreover, when used in combination with microorganisms, SMS-derived biochar can improve soil structure and increase nutrient cycling, thereby boosting nutrient availability and plant productivity [16,87]. Studies also suggest that combining biochar with mycorrhizal fungi can enhance soil structure, disease resistance, and plant growth [88–90], opening up new avenues for further research on SMS-derived biochar. This dynamic relationship between biochar and microbes supports sustainable agricultural practices, contributing to long-term improvements in soil health and ecosystem balance.

In wastewater treatment, biochar also provides a stable environment for pollutant-degrading bacteria, such as *Acinetobacter* [91], *Pseudomonas putida* [92], and *Rhodobacter capsulatus* [93], which help break down organic contaminants and facilitate nitrogen removal through nitrification and denitrification [94]. Combining SMS-derived biochar with various types of microbial life might extend the possibility of sustainable wastewater management by greatly enhancing the removal of heavy metals from wastewater, especially in the agricultural sector.

5. Applications of SMS-Derived Biochar in Modern Agricultural Systems

The application of SMS-derived biochar in modern agricultural systems offers numerous benefits (Figure 1), including enhanced soil amendment, improved nutrient retention, and mechanisms that promote plant growth and productivity [8,16,30,35]. Its utility extends to wastewater treatment in agricultural settings, where it aids in contaminant removal [21], and it plays a natural role in plant disease control by fostering effective microbial ecosystems [19]. Additionally, SMS-derived biochar significantly improves growth and productivity in mushroom production [14], making it a versatile tool for sustainable agriculture.

5.1. Soil Amendment

SMS-derived biochar plays a crucial role as a soil amendment in modern agricultural systems, particularly in enhancing soil fertility, improving water retention, and increasing organic matter content [7,8,35]. When incorporated into the soil, biochar improves the structure, porosity, and overall physicochemical and biological properties of the soil, facilitating better aeration and root penetration. Its high surface area and porous characteristics enable greater water retention, allowing soils to hold moisture for extended periods [7,95]. This capability mitigates irrigation needs and enhances drought resilience, which is particularly valuable in regions facing water scarcity [33].

Figure 1. Applications of biochar produced from SMS in modern agricultural practices.

SMS-derived biochar not only enhances soil physical properties but also plays a crucial role in reducing nutrient leaching while enriching the soil with essential nutrients and organic matter. Its composition includes macronutrients like N, P, and K, which directly support plant growth and improve crop yields, alongside micronutrients such as Ca [13,16]. Research by Lou et al. [35] highlighted that SMS-derived biochar exhibits excellent nutrient retention efficiency, as confirmed by leaching tests. The organic matter in this biochar encourages beneficial microbial activity, which is fundamental to enhancing soil fertility and nutrient cycling [96]. Furthermore, its application has been shown to improve the CEC of soils, increasing their ability to retain and make essential nutrients available to plants. This property is critical for maintaining long-term soil fertility and advancing sustainable agricultural practices [7]. Incorporating SMS-derived biochar as a soil amendment is a promising strategy for improving soil health and productivity. It offers an effective means of conserving nutrients without the need for additional materials, contributing significantly to sustainable agricultural systems and long-term ecological balance [35].

5.2. *Enhanced Nutrient Retention*

SMS-derived biochar significantly enhances nutrient retention in soils, providing a mechanism that improves the availability of essential nutrients to plants [7]. The porous structure of biochar increases the surface area, allowing it to adsorb and retain essential macronutrients and micronutrients. This adsorption occurs through various mechanisms, including electrostatic attraction, cation exchange, and the formation of surface complexes with nutrient ions. Moreover, the high CEC of SMS-derived biochar enables it to hold onto positively charged nutrient ions (cations) and release them slowly over time [82]. This slow-release characteristic is beneficial as it helps maintain a steady supply of nutrients in the soil, reducing the risk of nutrient leaching into water systems and promoting more efficient use of fertilizers, thus serving as a sustainable alternative to expensive chemical fertilizers [97,98]. Additionally, the organic matter in biochar enhances microbial activity, which further aids in the breakdown of organic compounds and the mineralization of nutrients, making them more accessible to plants. This combination of enhanced nutrient retention and gradual-release mechanisms not only improves soil fertility but also contributes to the sustainability of agricultural systems by reducing the need for synthetic fertilizers and minimizing environmental impacts associated with nutrient runoff [99]. Therefore, the incorporation of SMS-derived biochar into soil management practices represents a promising strategy for optimizing nutrient availability and promoting healthier and more productive crops [7,16].

5.3. Enhancing Plant Growth and Productivity

Applying biochar has generally been demonstrated to significantly improve plant growth and yield through several mechanisms, primarily involving nutrient retention, microbial activity, soil structure enhancement, and increased efficiency in plant nutrient uptake [100], making them available to plants over an extended period. This nutrient retention capacity reduces leaching and improves the efficiency of fertilizer use, contributing to higher crop yields. Moreover, the addition of biochar promotes beneficial microbial activity in the soil [100,101]. The organic matter present in SMS-derived biochar provides a habitat and food source for a diverse range of soil microorganisms. These microbes play a crucial role in nutrient cycling, breaking down organic materials, and enhancing nutrient availability to plants [99]. Several studies have shown that crops such as cauliflower, lettuce, and rapeseed exhibit increased growth and yield when planted in soil treated with SMS-derived biochar, demonstrating the positive impact of biochar on crop productivity [15,50,87]. In practical applications, experiments with various crops have yielded promising results. For example, research has indicated that incorporating SMS-derived biochar into the soil improved the growth of lettuce and other leafy vegetables by enhancing nutrient uptake and promoting healthier root systems [16,87,102]. In another study, experiments using *Pleurotus* SMS and its biochar in tissue culture of *Clinacanthus nutans* revealed promising outcomes, such as increased leaf numbers and shoot length, although the effects varied with concentration. While SMS supports plant development, SMS-derived biochar shows potential as a nutrient adsorbent. These applications highlight biochar's role in advancing green biotechnology and zero-waste initiatives by transforming agricultural waste into valuable resources [103] and supporting sustainable farming practices.

5.4. Wastewater Treatment in Agricultural and Environmental Systems

SMS-derived biochar has shown great promise in wastewater treatment applications due to its excellent adsorption properties, making it effective for removing contaminants from industrial sectors to agricultural runoff [15]. The porous structure and high surface area of biochar allow it to adsorb a range of pollutants, including heavy metals and pesticide residues, which are common in agricultural wastewater [16,100]. Studies have demonstrated that SMS-derived biochar can significantly reduce concentrations of heavy metals like Pb, Cd, Cr, Cu, and Zn, effectively improving water quality and minimizing environmental impact [15,30,47,77,80,81,104]. Additionally, SMS-derived biochar enhances the immobilization of heavy metals and organic contaminants like pesticide residues through adsorption [7]. For example, recent research demonstrated its effectiveness in reducing pyrethroid pesticide levels, such as tetramethrin, beta-cypermethrin, and fenvalerate, in water samples (tap water, lake water, and river water), contributing to safer crop management and sustainability [105]. This simple and efficient method has practical potential for detecting and managing contaminants in various water sources. Beyond water treatment, the integration of SMS-derived biochar into agricultural practices supports a circular economy by converting agricultural byproducts into valuable resources [6,21].

These applications highlight the potential of SMS-derived biochar as a sustainable solution for improving water quality in agricultural landscapes while promoting environmental health.

5.5. Natural Role for Plant Disease Control

The application of SMS-derived biochar demonstrates a high positive correlation with plant growth, yield, and biochemical responses, making it a valuable tool in sustainable agriculture [16]. One of its most notable benefits is its dual ability to manage plant pathogens while enhancing soil health. By improving the soil's physical and chemical properties and fostering beneficial microorganisms, biochar supports plant defense mechanisms against diseases [19]. For example, biochars derived from citrus wood and greenhouse wastes have been shown to induce systemic resistance against *Botrytis cinerea* in various hosts, including *Lycopersicon esculentum* (tomato), *Capsicum annuum* L. cv. Maccabi (bell pepper), and *Fragaria ananassa* (strawberry). Similarly, citrus and pine wood biochars have been effective in resisting diseases caused by *Colletotrichum acutatum* in strawberries, *Leveillula taurica* in bell peppers, and *Phytophthora cactorum* in red maple trees [17,19,106]. Numerous studies have demonstrated that biochar applications can lead to healthier crops by suppressing soil-borne pests and pathogens [17,19,107,108]. Its porous structure provides habitats for beneficial microorganisms, such as rhizobia, mycorrhizal fungi, actinomycetes, diazotrophic bacteria (plant-growth-promoting bacteria (PGPB)), and other soil microbiomes, which form symbiotic relationships with plant roots. These microbes contribute to nutrient mineralization, hormone production, and competitive inhibition of harmful pathogens, enhancing overall plant resilience [17,109]. Moreover, biochar's ability to boost microbial biomass and alter microbial community structures further supports diverse microbial activity by modifying soil physicochemical properties, ultimately benefiting plant growth and sustainability [110,111]. Research also indicates that using biochar at concentrations of 1% and 2% can significantly reduce the incidence of diseases in tomatoes, such as *Fusarium* wilt, thereby enhancing plant growth and productivity [112]. Additionally, biochar inhibits plant pathogenicity by enhancing resistance, increasing nutrient availability, and detoxifying pollutants, further reducing the stressful environment for plants and enhancing resistance to pests [113]. These benefits suggest the possibility that SMS-derived biochar not only improves soil fertility and structure but also offers a strategic eco-friendly solution for managing soil-borne pests and diseases. By integrating biochar into agricultural systems, farmers can enhance productivity while promoting sustainability.

5.6. Improving the Growth and Productivity in Mushroom Production

The incorporation of SMS-derived biochar into mushroom production substrates has demonstrated significant potential in enhancing the mycelial growth and yield of various mushroom species, especially oyster mushrooms like *P. ostreatus* and *P. pulmonarius* [14,20,36,114]. When added to growing media, biochar improves the substrate's aeration and water-holding capacity, creating a more conducive environment for mycelium development. This increased porosity facilitates better oxygen flow and moisture retention, both of which are critical for optimal mycelial growth [14,20]. As a result, mushroom farmers may have the chance to obtain higher-quality fruiting bodies and increased yields compared to traditional substrates that do not contain biochar. Moreover, SMS-derived biochar is rich in essential nutrients, trace elements, and carbon, which enrich the growing medium and can help the mycelium grow more effectively and provide the mushrooms with a consistent supply of nutrients [20].

Numerous studies have shown that incorporating SMS-derived biochar into the substrate enhances nutrient availability, leading to faster colonization and increased fruiting.

For instance, research by Hu et al. [14] indicated that using SMS-derived biochar resulted in 20 to 25% higher yields and reduced fruiting times by 4 to 6 days compared to conventional SMS. Similarly, other studies have reported that adding biochar to the growing substrate for oyster and button mushrooms leads to increased productivity and extended harvesting cycles [14,20,114–117]. Recent research, such as that by Kansara et al. [118], explored SMS-derived biochar produced from species like shiitake (*L. edodes*), blue oyster (*P. ostreatus*), and lion's mane (*H. erinaceus*). Trials with biochar at different mass loadings revealed notable improvements, particularly for lion's mane, with biochar concentrations around 5% significantly boosting yields. The findings highlight reduced colonization times and increased fruit body production, demonstrating biochar's role in enhancing substrate efficiency. Additionally, biochar stability slows substrate degradation and minimizes nutrient loss, ensuring sustained nutrient availability and promoting long-term cultivation efficiency [14,20]. These benefits, combined with the environmental advantages of waste heat recovery and circular waste-to-biochar processes, make SMS-derived biochar a sustainable choice for mushroom farming. Its adoption supports both commercial and small-scale growers, aligning with environmentally friendly cultivation practices in contemporary agricultural systems.

6. Comparison of SMS-Derived Biochar to Other Biochars

6.1. Performance Comparison of SMS-Derived Biochar to Other Biochars

The comparison between SMS-derived biochar and other biochar highlights the distinctive properties and advantages of SMS-derived biochar (Table 4). These unique characteristics make it a superior option in various agricultural and environmental applications. In terms of superior chemical and physical properties, SMS-derived biochar boasts a wide pH range (6.5–12.3), catering to diverse soil needs, whereas many other biochars, like those from acacia or bamboo wood, exhibit narrower pH ranges (e.g., 5.2–8.83). The C/N ratio of SMS-derived biochar (26.2–85.2) demonstrates balanced carbon and nitrogen content, promoting sustained microbial activity and nutrient cycling. In contrast, conventional options such as bagasse (C/N ~118) may limit microbial functionality due to excessive carbon.

The organic carbon content of SMS-derived biochar (37.5–70.7%) aligns with or surpasses that of traditional biochars, contributing to long-term soil carbon sequestration. Additionally, it boasts a higher nutrient density, with N (0.3–7.2%), P (0.03–3.8%), and K (0.08–1.4%) present in greater diversity and quantity than in biochars such as corn cob (1.2–4.3% N, 0.31% P, and 1.8% K), acacia trees (0.6–1% N, 0.1–1.14% P, and 0.71% K), or bagasse (0.6% N, 0.08% P, and 0.43% K). These macronutrients, paired with substantial micronutrient levels (e.g., up to 13.4% Ca, 0.5–1.7% Mg, and 0.04–1.6% S), create a superior amendment for enhancing soil fertility, often absent in some biochars from traditional feedstocks. Notably, its metal content, including iron (1500–24,000 mg·kg^{-1}), manganese (10–1200 mg·kg^{-1}), and zinc (84–1000 mg·kg^{-1}), facilitates enhanced plant nutrition. These values significantly outpace those of other biochars, such as bamboo wood or rice husk, which have far lower micronutrient concentrations.

In terms of surface area and porosity, SMS-derived biochar exhibits superior surface area (SSA: 3.1–1095 m^2 g^{-1}) and porosity, enabling improved water retention, aeration, and pollutant adsorption. This surpasses many other biochars, like coconut shell (SSA: ~613–626.8 m^2 g^{-1}) or corn cob (SSA: ~53.7–56.9 m^2 g^{-1}), which have limited adsorption capacities. The pore size distribution of SMS-derived biochar (0.9–50 nm) optimally supports microbial colonization, enhancing soil health and plant growth.

Table 4. The performance comparison of SMS-derived biochar to other biochars, via information modified and summarized from Leng et al. [83], Jatuwong et al. [88], Bolan et al. [99], Tomczyk et al. [119], Ravindiran et al. [120], Wijitkosum, [121], and Ayaz et al. [122].

Biochar Types	pH	C/N Ratio	OC (%)	N (%)	P (%)	K (%)	Ca (%)	Mg (%)	S (%)	Fe $mg\cdot kg^{-1}$	Mn $mg\cdot kg^{-1}$	Zn $mg\cdot kg^{-1}$	Cu $mg\cdot kg^{-1}$	SSA ($m^2\cdot g^{-1}$)	PS (nm)
SMS-derived biochars	6.5–12.3	26.2–85.2	37.5–70.7	0.44–2.7	0.03–3.8	0.08–1.4	0.4–13.4	0.5–1.7	0.04–1.6	1500–24,000	10–1200	84–1000	10	3.1–1095	0.9–50
Acacia tree	-	5.68–86	51.6–56.8	0.6–1	0.1–1.14	0.71	0.27–0.39	0.001–0.24	-	-	-	-	-	-	-
Apple wood chips	-	117.5–122.7	67–73.6	0.57–0.6	0.00400.18	0.6	2.42	0.32	-	5745	91.5	37.3	9.9	-	-
Bagasse	6.6–9.7	118	71	0.6	0.08	0.43	1.2	0.21	0.03	4800	300	400	15	259	-
Bamboo wood	5.2–8.83	95	76	0.8	0.44	3.5	0.05	0.14	0.08	70	40	6400	15	29.69–369.59	-
Castor stalk	-	16.3–43.8	61.3–78.4	1.4–4.8	0.54–2.98	0.4–3.48	0.8–8.89	0.43–1.42	-	210–4332	35–294	27.2–236	6.8–55	-	-
Cashew wood residues	-	-	-	0.9	0.01	0.13	-	-	-	185	32.3	18.45	10.2	-	-
Chicken manure	8.1–11.7	12.4–13.2	23.7–34.6	1.8–2.8	1.9–3	4.2–6	2–3	2.1–3.8	-	-	-	-	-	-	-
Cocao shell	-	21.7	36.8	1.7	0.59	4.05	2.92	1.47	-	-	-	-	-	-	-
Coconut shell	-	256.7	77	0.3	0.7	0.71	0.04	0.05	0.02	100	20	10	15	613–626.8	0.42–2.85
Coffee waste	9.7–10.05	112.9	79	0.7	0.03	0.35	0.4	0.08	0.025	150	40	45	15	-	-
Conocarpus wastes	-	91.7	64.2	0.7	0.084	0.038	4.34	0.343	2.28	-	-	-	-	-	-
Corn cob	7.7–10.1	18.6–65.9	79.1–80	1.2–4.3	0.31	1.8	0.14	0.16	0.11	800	50	80	20	53.7–56.9	59.66
Corn stover	7.2–10	50.9–53.7	50.9–69.8	1–1.3	0.14–0.23	1.71–2.88	0.03–10.65	0.10–0.59	0.07	-	142	132	963	3.1–712	-
Cotton stalk	-	42.3	71.9	1.7	1.22	0.4	4.33	1.12	-	4332	161	27.2	47.2	-	-
Cow manure	9.20	46.6–224	33.6–60.6	0.15–1.3	0.2–0.8	0.01–0.26	0.04–0.94	0.03–0.4	0.11	376	137	162	-	2.5–8	-
Dairy manure	6.8–10.5	37.8–186	55.8–56.7	0.3–1.5	1.0–1.69	1.43–2.31	2.67–4.48	1.22–2.06	0.11–0.15	36,700–44,800	525–867	363–363	99–163	1.6–186.5	-
Digested dairy manure	-	247.5–250.9	57.7–59.4	0.23–0.24	0.65–0.83	1.49–1.66	2.65–2.67	0.85–0.97	0.27–0.29	1656–2356	145–191	131–200	-	-	-
Eucalypt green waste	5.9–9.7	203.8–238.1	61.9–81.5	0.26–0.4	0.005–0.23	0.03–1.1	0.05–4.3	0.04–1.04	0.014–0.08	30–7000	180–1400	20–200	5.0–12	-	-
Fall yard waste	-	55.2	60.7	1.1	0.21	1.08	5.46	0.36	0.1	1504	555	70	-	-	-
Food waste	-	14.2	52.4	3.7	0.05	1.46	5.17	0.53	0.08	4431	179	39	-	-	-
Grass clippings	-	10.9	53.5	4.9	1.2	6.13	2.06	0.63	0.63	1557	360	150	-	-	-
Indian pine		65.7	72.3	1.1	0.11	0.04	0.17	0.07	0.63	-	-	-	-	-	-
Lantana weed	-	129.8	77.9	0.6	1.33	0.06	0.12	0.1	0.08	-	-	-	-	-	-
Lemon peel	-	28.7	66	2.3	0.27	2.9	2.1	2.6	0.06	90	30	20	25	-	-
Macadamia shell	-	195	78	0.4	0.24	2.19	0.37	0.17	-	1211	-	-	-	-	-
Oil palm fruit	-	46	69	1.5	0.51	3.02	1.4	0.64	0.14	200	200	120	55	-	-
Pearl millet stover	10.6	26.9–48.5	53.3–59.1	1.1–2.2	0.05–1.6	2.24–2.95	0.04–1.47	0.19–1.06	-	-	-	-	-	-	-
Poultry litter	8.7–10.3	19.2	38.3	2	0.9	1	2.5	0.3	-	2695	265	238	57	1–50.9	-
Pruning fruit waste	10.4–10.8	86.4–135	67.5–77.8	0.5–0.9	0.23–0.27	1.39	2.5	2.87	0.005	0.033	0.008	0.01	0.09	92	-
Rice husk	7.5–9.2	76.7	46	0.6	0.3	0.85	0.13	0.17	0.009	200	450	300	20	51.39–525	31.63
Rice straw	6.7–10.8	38.8–73.8	44.3–46.6	0.6–1.2	0.019–0.092	2.41–2.82	0.058	0.2	0.25	-	-	-	-	772.3	2.18
Sewage sludge	-	15.7	26.6	1.7	0.52	6.57	0.64	-	-	22,100	450	1520	380	408	5.21
Sugar maple waste	-	266.7	80	0.3	0.02	0.32	0.5	0.06	0.09	49.7	368	23.9	5	-	-
Swine manure	7.8–11	26.3–30.5	54.9–57.9	1.8–2.2	0.98–1.55	1.62–3.53	2.03–2.89	1.57–2.13	0.02–0.04	-	-	-	-	4.9–15.9	-
Swine solids	8.4–9.5	14.7–17	44.1–51.5	2.6–3.5	3.89–5.9	1.78–2.57	3.91–6.15	2.44–3.69	0.80–0.85	48,400–74,800	1453–2440	3181–4981	1538–2446	-	-
Tamarind wood	-	112.9	86	0.7	0.04	0.53	1	0.06	0.03	5	5	10	4	-	-
Turkey litter	9.9	12–23.6	44.8–49.3	1.9–4.1	2.68–3.63	4.01–5.59	4.04–5.61	0.85–1.24	0.41–0.55	27,800–36,500	710–986	690–909	535–762	-	-
Wheat straw	6.7–11	46.8–50.4	55.4–60.8	1.1–1.3	0.012–0.041	2.06–3.22	0.030–0.047	0.01–0.02	-	-	-	-	-	4.5–246.2	0.975
Willow wood waste	-	118.7	47.5	0.4	-	-	-	-	0.19	0.05	110	83.5	2.55	11.4–840.6	0.545

SMS-derived biochar also excels in versatility and environmental remediation. Its ability to adsorb heavy metals and pollutants surpasses that of traditional biochars due to its higher nutrient density and structural properties [100]. Moreover, studies demonstrate its exceptional performance in promoting mycelial growth and improving mushroom yields when incorporated into cultivation substrates [14,20,36,114], a feature unmatched by other biochars. While other biochars have merits, SMS-derived biochar outperforms in chemical richness, structural attributes, and versatility.

Another crucial aspect of its enhanced performance is the presence of nitrogen (N)-doped biochar, which has received significant attention due to its ability to modify surface morphology and properties after high-temperature pyrolysis [123]. These N-functionalized biochars improve cation exchange capacity, surface reactivity, and adsorption efficiency, making them essential for soil fertility enhancement, pollutant removal, and microbial interactions [124,125]. Notably, SMS from shiitake mushroom cultivation has been used as a carbon precursor to produce nitrogen-doped activated biochar, which has demonstrated efficiency in removing reactive orange-16 azo dye from water and treating contaminants in synthetic effluents and real sewage water. This highlights SMS as a valuable waste resource for activated carbon production, with N-doping significantly boosting adsorption performance [126]. However, advanced analytical techniques like Raman spectroscopy, X-ray photoelectron spectroscopy (XPS), nuclear magnetic resonance (NMR), thermogravimetric analysis (TGA), Brunauer–Emmett–Teller (BET) analysis, Fourier-transform infrared spectroscopy (FTIR), and scanning electron microscopy (SEM), among others, are crucial for identifying important N-functional groups (such as pyridinic-N, pyrrolic-N, graphitic-N, and oxidized-N) for the purpose of contributing to the characterization of these features. These functional groups play a critical role in nutrient retention, heavy metal adsorption, and catalytic activity [126,127]. With many applications in environmental management and sustainable agriculture, SMS-derived biochar is positioned as a high-end option for contemporary methods that promote ecological sustainability and productivity.

6.2. Economic Costs, Impact Analysis, and Environmental Effects

A comparison of the economic costs and environmental impacts of SMS-derived biochar and traditional biochars highlights significant advantages of SMS-derived biochar in terms of cost-effectiveness and sustainability (Table 5). SMS-derived biochar, offers a cost-effective alternative to biochars produced from conventional biomass feedstocks such as hardwoods, agricultural residues, and tree branches. This is largely due to the abundance and low cost of SMS, a byproduct of mushroom cultivation, which eliminates the need for expensive raw materials. For instance, SMS-derived biochar costs approximately \$117.5/ton in Chongqing, China [14], significantly lower than traditional biochars, such as those derived from bamboo (\$7000 in Alabama, USA) or switchgrass (\$5490 in the USA). The economic disparity becomes even more apparent when comparing SMS-derived biochar to biochars made from virgin wood feedstocks (\$17,800 in Massachusetts, USA) or tree branches (\$11,000 in Kansas, USA). Even moderately priced biochars, such as those from coconut shells (\$800 in the USA) or sewage sludge (\$560–1000 in the USA) [128,129], do not match the affordability of SMS-derived biochar. This cost advantage makes SMS-derived biochar an attractive option for large-scale applications, particularly in agriculture and environmental management.

Table 5. A comparison between the unit cost of SMS-derived biochar and biochar made from various kinds of biomass in different locations using summarized information collected by Hu et al. [14], Alhashimi and Aktas, [128], and Zhang et al. [129].

Biomass Types	Locations	$/Ton
SMS	Chongqing, China	117.5
Bamboo	Alabama, USA	7000
Chicken manure	Republic of Korea	1300
Coconut shell	USA	800
Coppiced hardwoods	UK	1600
Corn debris, manure, and forestry debris	Idaho, USA	1500–2000
Hardwood	Australia	2300
Mixed hardwood and softwood biochar	Central Canada	1000
Oiltea camellia shell	USA	670
Pinewood	Missouri, USA	900–1900
Pine tree-derived organic biomass	Oregon, USA	8300
Sewage sludge	USA	560–1000
Softwood chips	California, USA	3500
Various timbers mixed with softwoods	Vermont, USA	7200
Switchgrass	USA	5490
Tree branches	Kansas, USA	11,000
Virgin wood feedstock	Massachusetts, USA	17,800
Water oak wood	USA	770

The processing of SMS into organic fertilizers represents a common method of waste management, with value for its potential economic benefits [121]. A critical metric for assessing the practical value of recycled SMS is its input–output ratio compared to that of traditional organic fertilizers. According to Hu et al. [14], the production costs of SMS-derived biochar and organic fertilizers encompass four main components, i.e., facilities, energy, transportation, and labor, while the outputs are measured in terms of increased crop yields and cost savings. For a production volume of 1000 kg, the economic analysis reveals that the production cost of SMS-derived biochar is approximately $117.50, with potential economic benefits of around $414.80. In contrast, the cost of organic fertilizers is about $63.90, yielding economic benefits of $193.30. Although SMS costs are roughly double that of organic fertilizers, the output-to-input ratio for SMS-derived biochar stands at 4.53, compared to 4.02 for organic fertilizers, resulting in a total benefit for SMS-derived biochar that exceeds organic fertilizers by $221.50.

According to a case study by Zhang et al. [130], China generates around 15 million tons of SMS annually from edible mushroom production. If SMS-derived biochar manufacturing techniques were adopted nationally, the value added could reach an impressive $3.32 billion beyond that of organic fertilizers. This increase in productivity is associated with the reduced consumption of ecological resources and lower carbon emissions per agricultural product unit. Additionally, the conversion of SMS to SMS-derived biochar does not require fermentation, minimizing environmental risks and saving both time and space [14]. This process aligns with the principles of clean production, making it a more sustainable option than traditional organic fertilizers.

When examining the budget of biochar production enterprises, the baseline costs for producing 1 Mg (1000 kg) of biochar are approximately $717.76, excluding uncertainties in various parameters. The economic impacts of biochar production suggest an output-to-input ratio ranging from 2.01 to 2.85 (mean: 2.43) [131]. Notably, SMS-derived biochar production boasts variable costs that are six times lower than those of other biochar sources and offers a higher output-to-input ratio of 1.86 times. These findings highlight the economic viability and environmental benefits of utilizing SMS as a feedstock for SMS-derived biochar, promoting sustainable agricultural practices while effectively managing waste and reducing ecological impacts.

Beyond cost, SMS-derived biochar also excels in its environmental benefits. Utilizing SMS for biochar production addresses the waste management challenges associated with spent substrate disposal, effectively reducing the environmental footprint of mushroom farming. This conversion process recycles SMS into a valuable product that enhances soil health, improves nutrient retention, and supports pollution remediation [7,15]. Additionally, the multifunctionality of SMS-derived biochar further strengthens its environmental appeal. Its uses in agriculture, like improving soil fertility and supporting environmentally friendly farming practices, help ecological sustainability in the long term [7]. All things considered, SMS-derived biochar is notable for its minimal cost of production, effective waste management, and positive environmental impacts. By addressing both economic and ecological concerns, SMS-derived biochar presents itself as a superior alternative to traditional biochar, aligning with the goals of sustainable agriculture and innovative environmental practices.

7. Biochar Regeneration and Its Possibilities for Long-Term Sustainability

Biochar is widely regarded as being cost-effective because of its simple production processes and inexpensive feedstock. One significant advantage is its regenerability, which significantly reduces operational costs and allows for multiple reuse cycles in wastewater management [132]. Chemical regeneration has been utilized frequently to preserve the adsorption capability of biochar, providing a long-term sustainable solution. These techniques involve action treatment with salts, metal oxides, acids, and alkalis to enhance biochar's effectiveness [133]. However, challenges such as secondary contamination and varying regeneration efficiency, depending on adsorption mechanisms and pore structure, must be addressed. Various regeneration techniques have been developed, which are categorized into desorption (including thermal methods, such as steam, inert gas, hot water, and microwave, as well as non-thermal methods like chemical-based substances, surfactants, and supercritical fluids) and decomposition (including oxidation, electrochemical treatment, ultrasound, and microbiological processes) [132]. While thermal methods effectively break down pollutants, they are energy-intensive and costly. In contrast, non-thermal techniques, particularly chemical regeneration, offer a more sustainable and cost-efficient approach by preserving biochar's structure for repeated use [134,135]. This report demonstrates the necessity of more research to optimize regeneration effectiveness and support the use of biochar in sustainable wastewater management.

8. Potential Benefits Beyond Agriculture

Biochar offers numerous potential benefits beyond agriculture, including its use in animal husbandry for improving livestock health and reducing odors, enhancing carbon sequestration to combat climate change, and innovative applications in environmental management, such as wastewater treatment and pollution mitigation. Furthermore, its structural properties lend themselves well to construction and infrastructure projects, making it a versatile material in sustainable development initiatives (Figure 2) [95]. These diverse applications demonstrate the flexibility of biochar and create opportunities for SMS-derived biochar to be utilized beyond agricultural fields, extending its impact across multiple sectors.

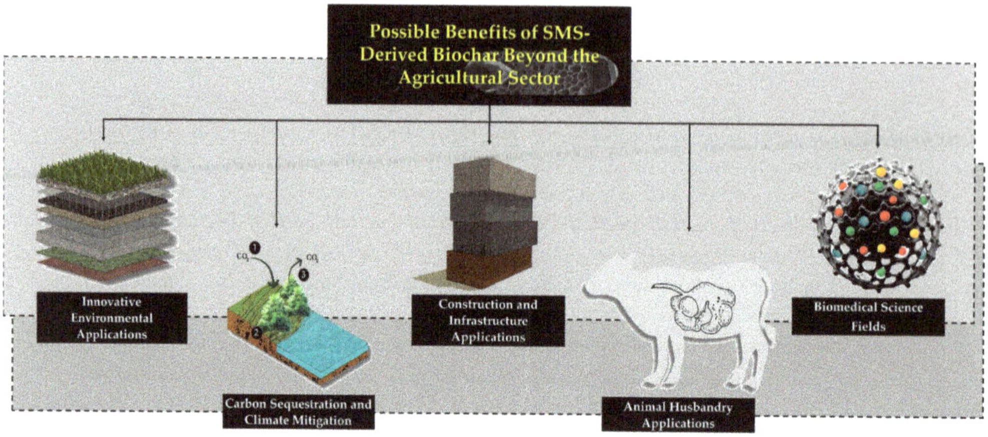

Figure 2. Potential advantages of biochar derived from SMS beyond agriculture.

8.1. Innovative Environmental Applications

A carbon-rich product produced by pyrolyzing organic materials, biochar has been proposed as the perfect substrate additive and offers creative uses in urban infrastructure to enhance the quality of runoff from green roofs. Its effective nutrient sorption capacity and ability to promote plant growth make it a perfect choice for urban gardening projects. By enhancing soil aeration and nutrient availability, biochar fosters healthier plant ecosystems in urban environments [136,137]. Moreover, incorporating biochar into city planning not only supports sustainable urban agriculture but also contributes to climate resilience, aligning with broader environmental management goals [138]. Research indicates that biochar derived from various sources, such as hardwoods, eucalyptus, mixed conifer wood, green waste, sewage sludge, coconut shell, and waste wood, can enhance both aesthetic and ecological outcomes in urban settings [136,139–144]. This presents another strategy for the potential of SMS-derived biochar as a sustainable urban infrastructure innovation, especially for green roofs.

8.2. Carbon Sequestration and Climate Mitigation

SMS-derived biochar plays a crucial role in carbon sequestration and climate mitigation by capturing atmospheric carbon and reducing overall greenhouse gas emissions [7,8,12]. When applied to soils, biochar enhances carbon storage and delivers numerous productivity benefits, including reduced bulk density, improved water retention, increased nutrient availability, stabilization of organic matter, and enhanced microbial activity, along with the ability to sequester heavy metals. Its stable carbon structure inhibits the rapid breakdown of organic matter, a primary contributor to carbon release, while also increasing phosphorus availability in highly weathered soils [11,100,145]. Converting local feedstocks and agricultural waste to biochar is particularly beneficial for smallholder farming, tree nurseries, and specialty crop management as it promotes soil health and fertility, leading to greater plant growth and further carbon absorption [145]. Numerous studies indicate that incorporating biochar into agricultural practices significantly reduces carbon emissions, positioning it as an effective strategy for climate change mitigation [7,10,146]. Utilizing SMS for biochar production not only supports agricultural sustainability but also contributes to global climate action efforts.

8.3. Construction and Infrastructure Applications

Incorporating biochar into construction materials is increasingly recognized for its potential to enhance building performance and sustainability [137]. Numerous studies show that adding biochar to building materials, such as bricks, concrete, and cement, can improve thermal insulation, reduce overall weight, and boost energy efficiency in buildings. This eco-friendly approach not only repurposes agricultural byproducts but also aligns with the construction industry's transition toward sustainable practices [137,147,148]. This approach suggests the integration of SMS-derived biochar in building practices, leading to the creation of more environmentally friendly structures that also advance innovative material science.

8.4. Animal Husbandry Applications

Several studies have shown that biochar supplementation in animal husbandry, like cows, pigs, poultry, and fish, can significantly enhance animal growth, improve hematological profiles, and boost the yield of meat, milk, and eggs [95,149,150]. Additionally, biochar has been shown to reduce methane emissions in ruminants, a crucial benefit for minimizing livestock's environmental impact. When mixed into animal diets, biochar supports nutrient absorption, aids in digestion, and removes contamination, all of which help the overall health and growth of the animal. Additionally, biochar use in livestock bedding or manure management reduces odors and pathogen presence, fostering a healthier environment [149]. Crucially, biochar in animal systems also helps in the recovery of essential nutrients from waste and raises the value of fertilizer made from manure, which may subsequently be recycled into agricultural soils, reducing negative impacts on the environment and promoting sustainable nutrient cycles [151]. This strategy demonstrates the additional benefits that SMS-derived biochar might offer for future sustainable animal husbandry.

8.5. Biomedical Science Fields

Beyond agricultural activity, N-doped biochar has received significant attention in biomedical applications due to its biocompatibility, antimicrobial properties, and photothermal efficiency [152]. It demonstrates superior antimicrobial activity against pathogens like *Escherichia coli* and Gram-positive *Staphylococcus aureus*, positioning it as a potential alternative to traditional antibiotics [153]. Additionally, N-doped carbon-based materials serve as effective photosensitizing agents in photodynamic therapy (PDT), generating reactive oxygen species (ROS) to target and destroy cancer cells [154]. These materials are also used in drug delivery systems, showcasing their multifunctional potential in medical science [152]. This offers an informative opportunity to explore SMS-derived biochar in advanced biomedical applications.

9. Challenges and Limitations of SMS-Derived Biochar Utilization

Carbonization, particularly through pyrolysis, has gained significant attention in recent years for its potential applications in agricultural and environmental contexts [13]. SMS is a distinctive feedstock for carbonization due to its rich content of amino acids and small organic molecules, which gives carbonized SMS, or SMS-derived biochar, unique properties [35]. SMS-derived biochar has shown promising potential, with applications as a fertilizer and even as a growth medium for further mushroom cultivation [13,20,155]. However, despite advances in its use as an adsorbent and fertilizer, the role of carbonized SMS in soil improvement remains underexplored, and research into its use as biochar is still growing [13]. Additionally, biochar-induced dust emissions pose potential long-term health risks to users, particularly in agricultural applications. Developing alternative biochar forms, such as granules, pellets, dust-free types, and liquid biochar fertilizers, could offer more convenient, effective, and lower-risk solutions. Moreover, a study by Horn et al. [156] indicated that the excessive use of biochar will not only impact the environment but also affect the composition of soil matter, potentially affecting the soil's natural functions.

The limited studies available highlight the need to investigate the effects of different SMS types and pyrolysis temperatures on the biochar's characteristics and its adsorption capacity for various heavy metals [49]. Many studies have demonstrated that varying these parameters may influence SMS-derived biochar properties, especially its ability to retain essential nutrients and remove contaminants from the soil [12,20,49]. Furthermore, current methods for creating SMS-derived biochar often involve complex and expensive laboratory techniques, which restrict their feasibility for large-scale applications. For SMS-derived biochar to be commercially viable, studies are needed to streamline and optimize the pyrolysis of agricultural wastes like SMS to produce biochar with high adsorption capabilities and practical utility in soil amelioration [46]. This emerging research direction holds promise for further enhancing the ecological and economic value of SMS-derived biochar in sustainable agricultural systems.

10. Future Prospects and Research Opportunities

Future research on SMS-derived biochar opens up promising avenues for enhancing agricultural and environmental sustainability. One interesting application is in precision agriculture, where biochar can support soil improvements, controlled-release fertilizers, and advanced water management systems to boost resource efficiency and crop performance. By integrating biochar into modern soil management techniques, farmers can increase nutrient efficiency, reduce input waste, and decrease their use of synthetic fertilizers [8,16,35].

Improving SMS-derived biochar quality and functionality through modified production conditions, chemical enhancements, or blends with other materials is another promising approach, potentially increasing nutrient retention, stability, and pollutant removal efficiency. For example, pyrolyzing biochar at temperatures above 500 °C can improve its capacity to handle multiple contaminants, while acid treatments may enhance pollutant absorption [157]. These modifications can expand the application of biochar beyond soil enhancement, making it a viable solution for water filtration, heavy metal remediation, and carbon sequestration.

Beyond technical advancements, policy support and regulatory frameworks are also crucial to promoting sustainable biochar practices. Government incentives, carbon credit programs, and waste management regulations can make biochar a more economically viable solution, furthering its role in carbon sequestration, waste recycling, soil health, and reducing greenhouse gas emissions in agriculture systems [74,146]. Ultimately, these advancements in SMS-derived biochar could transform it into a highly valuable tool for sustainable agricultural management and environmental conservation. Achieving these goals will require collaboration between researchers, industry leaders, and policymakers to drive innovation and implementation [1].

Importantly, growing biochar technology requires a comprehensive and cooperative strategy that integrates industry-driven innovations, scientific findings, and supportive requirements to optimize its potential. Continuous research, strategic partnerships, and well-defined production standards will be crucial in realizing the full potential of SMS-derived biochar. With sustained efforts in innovation and implementation, SMS-derived biochar can potentially become an important part of sustainable agriculture and environmental resilience and contribute to a circular economy.

11. Conclusions

The unique physicochemical properties of SMS, including its lignocellulosic composition, high nutritional profile, and organic matter richness, make it a perfect substrate for biochar production. SMS-derived biochar offers a sustainable solution to numerous challenges in modern agricultural systems. This extensive review discusses the multifaceted benefits of SMS-derived biochar, from its composition and production processes to its applications in soil enhancement, plant growth, water treatment, pest management, and mushroom cultivation. SMS-derived biochar demonstrates superior characteristics, such as increased porosity, nutrient retention, and carbon stability, making it highly effective in improving soil fertility, water retention, and microbial ecosystems and in sequestering carbon to mitigate climate change. Economically, SMS-derived biochar offers a viable alternative to other biochars due to the abundance and low cost of its precursor. Its integration into agricultural systems reduces dependence on synthetic fertilizers, enhancing resource efficiency for both smallholder and industrial farming. Despite its broad potential, challenges such as feedstock variability, diverse pyrolysis methods, and the need for long-term environmental impact assessments remain. Future research should focus on standardizing production methods, exploring innovative applications, studying regeneration techniques, and developing policies to integrate SMS-derived biochar into sustainable agricultural frameworks. Overall, SMS-derived biochar stands as a key innovation in sustainable agriculture systems, aligning with ecological conservation, resource efficiency, and sustainable development goals. It also provides a viable framework for promoting green agricultural practices while contributing to global environmental flexibility.

Author Contributions: Conceptualization, W.A. and S.L.; software, W.A.; validation, W.A., K.J., T.K., W.K., P.T., R.J. and S.L.; formal analysis, W.A., K.J. and S.L.; data curation, W.A., K.J., T.K., W.K., P.T., R.J. and S.L.; writing—original draft preparation, W.A., K.J. and R.J.; writing—review and editing, W.A., K.J., T.K., W.K., P.T., R.J., O.X. and S.L.; supervision, S.L.; project administration, W.A.; and funding acquisition, S.L. and T.K. All authors have read and agreed to the published version of the manuscript.

Funding: This work was supported by Chiang Mai University. The authors gratefully acknowledge the CMU Proactive Researcher program, Chiang Mai University [grant number 784/2567], Chiang Mai, Thailand.

Institutional Review Board Statement: Not applicable.

Informed Consent Statement: Not applicable.

Data Availability Statement: Data are contained within the article.

Acknowledgments: This research work was partially supported by Chiang Mai University, Chiang Mai, Thailand. The authors would like to express their gratitude to Kingkarn Rattanapong and Thanaporn Promjaidee from Yupparaj Wittayalai School for their valuable contributions in researching information and preparing some of the illustrations for this literature review. Special thanks are also extended to Praween Jinanukul for his outstanding work in preparing the graphic designs that greatly enhanced the quality of this manuscript.

Conflicts of Interest: The authors declare no conflicts of interest.

References

1. Baptista, F.; Almeida, M.; Paié-Ribeiro, J.; Barros, A.N.; Rodrigues, M. Unlocking the potential of spent mushroom substrate (SMS) for enhanced agricultural sustainability: From environmental benefits to poultry nutrition. *Life* **2023**, *3*, 1948. [CrossRef] [PubMed]
2. Velusami, B.; Jordan, S.N.; Curran, T.; Grogan, H. Fertiliser characteristics of stored spent mushroom substrate as a sustainable source of nutrients and organic matter for tillage, grassland and agricultural soils. *J. Agric. Food Res. Ir.* **2022**, *60*, 1–11. [CrossRef]

3. Arora, A. Top-10 Mushroom Producing Countries in the World. 2024. Available online: https://currentaffairs.adda247.com/top-10-mushroom-producing-countries-in-the-world/ (accessed on 26 October 2024).
4. Kwiatkowska, E.; Joniec, J. Effects of agricultural management of spent mushroom waste on phytotoxicity and microbiological transformations of C, P, and S in soil and their consequences for the greenhouse effect. *Int. J. Environ. Res. Public. Health* **2022**, *19*, 12915. [CrossRef] [PubMed]
5. Wilkinson, K. Recycling Spent Mushroom Substrate (SMS) for Fertilizer in a Circular Economy. Available online: https://www.horticulture.com.au/growers/help-your-business-grow/research-reports-publications-fact-sheets-and-more/mu21006/ (accessed on 26 October 2024).
6. Martín, C.; Zervakis, G.I.; Xiong, S.; Koutrotsios, G.; Strætkvern, K.O. Spent substrate from mushroom cultivation: Exploitation potential toward various applications and value-added products. *Bioengineered* **2023**, *14*, 2252138. [CrossRef] [PubMed]
7. Zhao, Z.; Ibrahim, M.M.; Wang, X.; Xing, S.; Heiling, M.; Hood-Nowotny, R.; Tong, C.; Mao, Y. Properties of Biochar Derived from Spent Mushroom Substrates. *BioResources* **2019**, *14*, 5254–5277. [CrossRef]
8. Xu, X.; Yuan, X.; Zhang, Q.; Wei, Q.; Liu, X.; Deng, W.; Wang, J.; Yang, W.; Deng, B.; Zhang, L. Biochar derived from spent mushroom substrate reduced N_2O emissions with lower water content but increased CH_4 emissions under flooded condition from fertilized soils in *Camellia oleifera* plantations. *Chemosphere* **2022**, *287*, 132110. [CrossRef]
9. Luo, J.; Chen, L. Status and development of spent mushroom substrate recycling: A review. *J. Air Waste Manag. Assoc.* **2024**, *12*, 843–860. [CrossRef] [PubMed]
10. Afshar, M.; Mofatteh, S. Biochar for a sustainable future: Environmentally friendly production and diverse applications. *Results Eng.* **2024**, *23*, 102433. [CrossRef]
11. Shoudho, K.N.; Khan, T.H.; Ara, U.R.; Khan, M.R.; Shawon, Z.B.Z.; Hoque, M.E. Biochar in global carbon cycle: Towards sustainable development goals. *Curr. Res. Green Sustain. Chem.* **2024**, *8*, 100409. [CrossRef]
12. Deng, B.; Shi, Y.; Zhang, L.; Fang, H.; Gao, Y.; Luo, L.; Feng, W.; Hu, X.; Wan, S.; Huang, W.; et al. Effects of spent mushroom substrate-derived biochar on soil CO_2 and N_2O emissions depend on pyrolysis temperature. *Chemosphere* **2020**, *246*, 125608. [CrossRef] [PubMed]
13. Chen, W.S.; Tsai, W.T.; Lin, Y.Q.; Tsai, C.H.; Chang, Y.T. Production of highly porous biochar materials from spent mushroom composts. *Horticulturae* **2022**, *8*, 46. [CrossRef]
14. Hu, W.; Di, Q.; Liang, T.; Liu, J.; Zhang, J. Effects of spent mushroom substrate biochar on growth of oyster mushroom (*Pleurotus ostreatus*). *Environ. Technol. Innov.* **2022**, *28*, 102729. [CrossRef]
15. Madzin, Z.; Zahidi, I.; Raghunandan, M.E.; Talei, A. Potential application of spent mushroom compost (SMC) biochar as low-cost filtration media in heavy metal removal from abandoned mining water: A review. *Int. J. Environ. Sci. Technol.* **2023**, *20*, 6989–7006. [CrossRef] [PubMed]
16. Širić, I.; Eid, E.M.; Taher, M.A.; El-Morsy, M.H.; Osman, H.E.; Kumar, P.; Adelodun, B.; Fayssal, S.A.; Mioč, B.; Andabaka, Z.; et al. Combined use of spent mushroom substrate biochar and PGPR improves growth, yield, and biochemical response of cauliflower (*Brassica oleracea* var. *botrytis*): A preliminary study on greenhouse cultivation. *Horticulturae* **2022**, *8*, 830. [CrossRef]
17. Bhatt, B.; Gupta, S.K.; Mukherjee, S.; Kumar, R. A comprehensive review on biochar against plant pathogens: Current state-of-the-art and future research perspectives. *Heliyon* **2024**, *10*, e37204. [CrossRef] [PubMed]
18. Kittimorakul, J.; Chaiyama, V.; Wattanakul, A. Effect of biochar derived from spent mushroom substrates (DOA Mushchar) for improving the production efficiency of mother culture and mother spawn of Lingzhi mushroom. *Songklanakarin J. Plant Sci.* **2022**, *9*, 32–38.
19. Dorjee, L.; Nishmitha, K.; Pattanayak, S.; Wangmu, T.; Meshram, S.; Chongtham, S.; Gogoi, R. Biochar: A comprehensive review on a natural approach to plant disease management. *J. Pure Appl. Microbiol.* **2024**, *18*, 29–45. [CrossRef]
20. Lam, S.S.; Lee, X.Y.; Nam, W.L.; Phang, X.Y.; Liew, R.K.; Yek, P.N.; Ho, Y.L.; Ma, N.L.; Rosli, M.H. Microwave vacuum pyrolysis conversion of waste mushroom substrate into biochar for use as growth medium in mushroom cultivation. *J. Chem. Technol. Biotechnol.* **2019**, *94*, 1406–1415. [CrossRef]
21. Darajeh, N.; Alizadeh, H.; Leung, D.W.; Rashidi Nodeh, H.; Rezania, S.; Farraji, H. Application of modified spent mushroom compost biochar (Smcb/fe) for nitrate removal from aqueous solution. *Toxics* **2021**, *9*, 277. [CrossRef] [PubMed]
22. Lehmann, J.; Joseph, S. (Eds.) *Biochar for Environmental Management: Science, Technology and Implementation*; Taylor & Francis: Milton Park, UK, 2024.
23. Ahlawat, O.P.; Sagar, M.P. *Management of Spent Mushroom Substrate Technical Bulletin*; National Research Centre for Mushroom, Indian Council of Agricultural Research: Solan, India, 2007.
24. Catal, S.; Peksen, A. Physical, chemical and biological properties of spent mushroom substrates of different mushroom species. *Acta Hortic.* **2020**, *1287*, 353–360. [CrossRef]
25. Corral-Bobadilla, M.; González-Marcos, A.; Vergara-González, E.P.; Alba-Elías, F. Bioremediation of waste water to remove heavy metals using the spent mushroom substrate of *Agaricus bisporus*. *Water* **2019**, *11*, 454. [CrossRef]

26. Carmo, C.O.; Rodrigues, M.D.S.; Silva, F.D.; Irineu, T.G.; Soares, A.C.F. Spent mushroom substrate of *Pleurotus ostreatus* Kmmer increases basil biomass and essential oil yield. *Rev. Caatinga* **2021**, *34*, 548–558. [CrossRef]
27. Leong, Y.K.; Ma, T.W.; Chang, J.S.; Yang, F.C. Recent advances and future directions on the valorization of spent mushroom substrate (SMS): A review. *Bioresour. Technol.* **2022**, *344*, 126157. [CrossRef] [PubMed]
28. Adi, M.R.M.; Izhar, T.N.T.; Ibrahim, N.M.; Abd Aziz, N.; Hadiyanto, H.; Matei, M. The effect of chemical treated spent mushroom substrate (SMS) on lignocellulosic content. *IOP Conf. Ser. Earth Environ. Sci.* **2023**, *1216*, 012015. [CrossRef]
29. Deshmukh, A.S. Spent mushroom substrate: A treasure of nutrients. *Asian J. Microbiol. Biotechnol. Environ. Sci.* **2019**, *21*, 1024–1027.
30. Chang, J.; Zhang, H.; Cheng, H.; Yan, Y.; Chang, M.; Cao, Y.; Huang, F.; Zhang, G.; Yan, M. Spent *Ganoderma lucidum* substrate derived biochar as a new bio-adsorbent for Pb^{2+}/Cd^{2+} removal in water. *Chemosphere* **2020**, *241*, 125121. [CrossRef]
31. Pintarič, M.; Štuhec, A.; Tratnik, E.; Langerholc, T. Spent Mushroom Substrate Improves Microbial Quantities and Enzymatic Activity in Soils of Different Farming Systems. *Microorganisms* **2024**, *12*, 1521. [CrossRef] [PubMed]
32. Chen, Y.; Sun, Z.; Su, Y.; Yang, J.; Li, M.; Hong, B.; Chen, G. Hydrochar Derived from Spent Mushroom Substrate Ameliorates Soil Properties and Nutrient Levels in Saline–Sodic Soil: An Incubation Study. *Sustainability* **2022**, *14*, 12958. [CrossRef]
33. Wu, H.; Xue, S.; Wang, L.; Chen, Y.; Xu, Y.; Xu, Y.; Chen, F. *Pleurotus pulmonarius* pretreated lignocellulosic feedstock as a sustainable precursor to produce biochar for efficient removal of NO_3^-. *ACS Sustain. Resour. Manag.* **2024**, *1*, 530–538. [CrossRef]
34. Li, S.; Tasnady, D. Biochar for soil carbon sequestration: Current knowledge, mechanisms, and future perspectives. *C J. Carbon Res.* **2023**, *9*, 67. [CrossRef]
35. Lou, Z.; Sun, Y.; Bian, S.; Ali Baig, S.; Hu, B.; Xu, X. Nutrient conservation during spent mushroom compost application using spent mushroom substrate derived biochar. *Chemosphere* **2017**, *169*, 23–31. [CrossRef]
36. Sangsuwan, P.; Detraksa, J.; Srimawong, P. Proximate analysis of the growth of organic grey oyster mushrooms on biochar from agricultural waste. *Pak. J. Phytopathol.* **2023**, *35*, 83–91. [CrossRef]
37. Lou, Z.; Zhu, J.; Wang, Z.; Baig, S.A.; Fang, L.; Hu, B.; Xu, X. Release characteristics and control of nitrogen, phosphate, organic matter from spent mushroom compost amended soil in a column experiment. *Process Saf. Environ. Prot.* **2015**, *98*, 417–423. [CrossRef]
38. Domínguez-Gutiérrez, M.; Gaitán-Hernández, R.; Moctezuma-Pérez, I.; Barois, I.; Domínguez, J. Composting and vermicomposting of spent mushroom substrate to produce organic fertilizer. *Emir. J. Food Agric.* **2022**, *34*, 220–228. [CrossRef]
39. Goglio, P.; Ponsioen, T.; Carrasco, J.; Milenkovi, I.; Kiwala, L.; Van Mierlo, K.; Helmes, R.; Tei, F.; Oosterkamp, E.; Pérez, M. An environmental assessment of *Agaricus bisporus* ((JE Lange) Imbach) mushroom production systems across Europe. *Eur. J. Agron.* **2024**, *155*, 127108. [CrossRef]
40. Gotore, O.; Itayama, T.; Dang, B.T.; Nguyen, T.D.; Ramaraj, R.; Osamu, N.; Shuji, T.; Maseda, H. Adsorption analysis of ciprofloxacin and delafloxacin onto the corn cob derived-biochar under different pyrolysis conditions. *Biomass Convers. Biorefin.* **2024**, *14*, 10373–10388. [CrossRef]
41. Zhang, Y.; Chen, P.; Liu, S.; Peng, P.; Min, M.; Cheng, Y.; Anderson, E.; Zhou, N.; Fan, L.; Liu, C.; et al. Effects of Feedstock Characteristics on Microwave-Assisted Pyrolysis—A Review. *Bioresour. Technol.* **2017**, *230*, 143–151. [CrossRef]
42. Pahnila, M.; Koskela, A.; Sulasalmi, P.; Fabritius, T. A review of pyrolysis technologies and the effect of process parameters on biocarbon properties. *Energies* **2023**, *16*, 6936. [CrossRef]
43. Armah, E.K.; Chetty, M.; Adedeji, J.A.; Estrice, D.E.; Mutsvene, B.; Singh, N.; Tshemese, Z. Biochar: Production, application and the future. In *Biochar-Productive Technologies, Properties and Applications*; IntechOpen: London, UK, 2022.
44. Ma, Y.; Wang, Q.; Sun, X.; Wang, X.; Su, W.; Song, N. A study on recycling of spent mushroom substrate to prepare chars and activated carbon. *BioResources* **2014**, *9*, 3939–3954. [CrossRef]
45. Xian, Y.; Wu, J.; Yang, G.; Liao, R.; Zhang, X.; Peng, H.; Yu, X.; Shen, F.; Li, L.; Wang, L. Adsorption characteristics of Cd (II) in aqueous solutions using spent mushroom substrate biochars produced at different pyrolysis temperatures. *RSC Adv.* **2018**, *8*, 28002–28012. [CrossRef]
46. Zhang, H.; Su, L.; Cheng, C.; Cheng, H.; Chang, M.; Liu, F.; Liu, N.; Oh, K. A new type of calcium-rich biochars derived from spent mushroom substrates and their efficient adsorption properties for cationic dyes. *Front. Bioeng. Biotechnol.* **2022**, *10*, 1007630. [CrossRef] [PubMed]
47. Zhang, G.; Liu, N.; Luo, Y.; Zhang, H.; Su, L.; Oh, K.; Cheng, H. Efficient removal of Cu (II), Zn (II), and Cd (II) from aqueous solutions by a mineral-rich biochar derived from a spent mushroom (*Agaricus bisporus*) substrate. *Materials* **2020**, *14*, 35. [CrossRef] [PubMed]
48. Grimm, A.; dos Reis, G.S.; Dinh, V.M.; Larsson, S.H.; Mikkola, J.P.; Lima, E.C.; Xiong, S. Hardwood spent mushroom substrate–based activated biochar as a sustainable bioresource for removal of emerging pollutants from wastewater. *Biomass Convers. Biorefin.* **2024**, *14*, 2293–2309. [CrossRef]

49. Jin, Y.; Zhang, M.; Jin, Z.; Wang, G.; Li, R.; Zhang, X.; Liu, X.; Qu, J.; Wang, H. Characterization of biochars derived from various spent mushroom substrates and evaluation of their adsorption performance of Cu (II) ions from aqueous solution. *Environ. Res.* **2021**, *196*, 110323. [CrossRef]
50. Zhang, M.; Jin, Z.; Wang, G.; Zhang, X.; Li, R.; Jin, Y.; Qu, J. Organic amendments prepared from spent mushroom substrate efficiently immobilize Cd and attenuate Cd uptake by rape seedlings. *Environ. Eng. Sci.* **2021**, *38*, 685–694. [CrossRef]
51. Sarfraz, R.; Li, S.; Yang, W.; Zhou, B.; Xing, S. Assessment of physicochemical and nutritional characteristics of waste mushroom substrate biochar under various pyrolysis temperatures and times. *Sustainability* **2019**, *11*, 277. [CrossRef]
52. Aboelela, D.; Saleh, H.; Attia, A.M.; Elhenawy, Y.; Majozi, T.; Bassyouni, M. Recent advances in biomass pyrolysis processes for bioenergy production: Optimization of operating conditions. *Sustainability* **2023**, *15*, 11238. [CrossRef]
53. Awad, M.I.; Makkawi, Y.; Hassan, N.M. Yield and energy modeling for biochar and bio-oil using pyrolysis temperature and biomass constituents. *ACS Omega* **2024**, *9*, 18654–18667. [CrossRef] [PubMed]
54. Jerzak, W.; Acha, E.; Li, B. Comprehensive review of biomass pyrolysis: Conventional and advanced technologies, reactor designs, product compositions and yields, and techno-economic analysis. *Energies* **2024**, *17*, 5082. [CrossRef]
55. Gryta, A.; Skic, K.; Adamczuk, A.; Skic, A.; Marciniak, M.; Józefaciuk, G.; Boguta, P. The importance of the targeted design of biochar physicochemical properties in microbial inoculation for improved agricultural productivity—A review. *Agriculture* **2023**, *14*, 37. [CrossRef]
56. Ndoung, O.C.N.; de Figueiredo, C.C.; Ramos, M.L.G. A scoping review on biochar-based fertilizers: Enrichment techniques and agro-environmental application. *Heliyon* **2021**, *7*, e08473. [CrossRef] [PubMed]
57. Enaime, G.; Baçaoui, A.; Yaacoubi, A.; Lübken, M. Biochar for wastewater treatment—Conversion technologies and applications. *Appl. Sci.* **2020**, *10*, 3492. [CrossRef]
58. Weber, K.; Quicker, P. Properties of biochar. *Fuel* **2018**, *217*, 240–261. [CrossRef]
59. Al Masud, M.A.; Shin, W.S.; Sarker, A.; Septian, A.; Das, K.; Deepo, D.M.; Iqbal, M.A.; Islam, A.R.M.T.; Malafaia, G. A critical review of sustainable application of biochar for green remediation: Research uncertainty and future directions. *Sci. Total Environ.* **2023**, *904*, 166813. [CrossRef]
60. Joseph, S.; Taylor, P.; Cowie, A. Basic Principles of Biochar Production. Available online: https://biochar.international/guides/basic-principles-of-biochar-production/#introduction (accessed on 16 November 2024).
61. Blenis, N.; Hue, N.; Maaz, T.M.; Kantar, M. Biochar production, modification, and its uses in soil remediation: A review. *Sustainability* **2023**, *15*, 3442. [CrossRef]
62. Tan, X.; Liu, Y.; Gu, Y.; Xu, Y.; Zeng, G.; Hu, X.; Liu, S.; Wang, X.; Liu, S.; Li, J. Biochar-based nano-composites for the decontamination of wastewater: A review. *Bioresour. Technol.* **2016**, *212*, 318–333. [CrossRef] [PubMed]
63. Gong, J.; Lin, H.; Antonietti, M.; Yuan, J. Nitrogen-doped porous carbon nanosheets derived from poly (ionic liquid)s: Hierarchical pore structures for efficient CO_2 capture and dye removal. *J. Mater. Chem. A* **2016**, *4*, 7313–7321. [CrossRef]
64. Pan, X.; Gu, Z.; Chen, W.; Li, Q. Preparation of biochar and biochar composites and their application in a Fenton-like process for wastewater decontamination: A review. *Sci. Total Environ.* **2021**, *754*, 142104. [CrossRef]
65. Wang, J.; Wang, S. Preparation, modification and environmental application of biochar: A review. *J. Clean. Prod.* **2019**, *227*, 1002–1022. [CrossRef]
66. Zhang, X.; Zhang, S.; Yang, H.; Shi, T.; Chen, Y.; Chen, H. Influence of NH_3/CO_2 modification on the characteristic of biochar and the CO_2 capture. *Bioenergy Res.* **2013**, *6*, 1147–1153.
67. Azeta, O.; Ayeni, A.O.; Agboola, O.; Elehinafe, F.B. A review on the sustainable energy generation from the pyrolysis of coconut biomass. *Sci. Afr.* **2021**, *13*, e00909. [CrossRef]
68. Zhang, C.; Ji, Y.; Li, C.; Zhang, Y.; Sun, S.; Xu, Y.; Jiang, L.; Wu, C. The application of biochar for CO_2 capture: Influence of biochar preparation and CO_2 capture reactors. *Ind. Eng. Chem. Res.* **2023**, *62*, 17168–17181. [CrossRef]
69. Petrov, P.; Nurmi, K.; Miles, T.; Joseph, S.; Nurmi, C.; Siemons, R.; Lee, N.H.; Echo Tech Carbon Corp. Solving for CO2-Climate Trend Setters. Available online: https://africacarbonmarkets.org/wp-content/uploads/2023/05/Detailed-Project-Design-.pdf (accessed on 26 October 2024).
70. Atallah, E.; Zeaiter, J.; Ahmad, M.N.; Leahy, J.J.; Kwapinski, W. Hydrothermal carbonization of spent mushroom compost waste compared against torrefaction and pyrolysis. *Fuel Process. Technol.* **2021**, *216*, 106795. [CrossRef]
71. Syguła, E.; Koziel, J.A.; Białowiec, A. Proof-of-concept of spent mushrooms compost torrefaction—Studying the process kinetics and the influence of temperature and duration on the calorific value of the produced biocoal. *Energies* **2019**, *12*, 3060. [CrossRef]
72. Kamal Baharin, N.S.; Koesoemadinata, V.C.; Nakamura, S.; Azman, N.F.; Muhammad Yuzir, M.A.; Md Akhir, F.N.; Iwamoto, K.; Yahya, W.J.; Othman, N.; Ida, T.; et al. Production of Bio-Coke from spent mushroom substrate for a sustainable solid fuel. *Biomass Convers. Biorefin.* **2020**, *12*, 4095–4104. [CrossRef]
73. Sewu, D.D.; Jung, H.; Kim, S.S.; Lee, D.S.; Woo, S.H. Decolorization of cationic and anionic dye-laden wastewater by steam-activated biochar produced at an industrial-scale from spent mushroom substrate. *Bioresour. Technol.* **2019**, *277*, 77–86. [CrossRef] [PubMed]

74. Pourhashem, G.; Hung, S.Y.; Medlock, K.B.; Masiello, C.A. Policy support for biochar: Review and recommendations. *GCB Bioenergy* **2019**, *11*, 364–380. [CrossRef]
75. Nam, W.L.; Phang, X.Y.; Su, M.H.; Liew, R.K.; Ma, N.L.; Rosli, M.H.N.B.; Lam, S.S. Production of bio-fertilizer from microwave vacuum pyrolysis of palm kernel shell for cultivation of oyster mushroom (*Pleurotus ostreatus*). *Sci. Total Environ.* **2018**, *624*, 9–16. [CrossRef] [PubMed]
76. Ibrahim, E.A.; El-Sherbini, M.A.; Selim, E.M.M. Effects of biochar on soil properties, heavy metal availability and uptake, and growth of summer squash grown in metal-contaminated soil. *Sci. Hortic.* **2022**, *301*, 111097. [CrossRef]
77. Liu, M.; Liu, X.; Wu, Z.; Zhang, Y.; Meng, Q.; Yan, L. Sulfur-modified *Pleurotus ostreatus* spent substrate biochar enhances the removal of cadmium in aqueous solution: Characterization, performance, mechanism. *J. Environ. Manag.* **2022**, *322*, 115900. [CrossRef]
78. Lin, J.C.; Cheng, A.C.; Shiu, Y.L.; Wong, Y.C.; Yeh, S.P.; Simangunsong, T.; Liu, C.H. Using the biochar produced from spend mushroom substrate to improve the environmental condition of aquaculture pond. *Aquac. Res.* **2021**, *52*, 3532–3539. [CrossRef]
79. Jiang, C.; Huang, S.; Jiang, Y.; Li, Y.; Miao, T.; Jin, Y.; Qu, J.; Liu, X.; Wang, W. Effects of aging processes on spent mushroom Substrate-Derived Biochar: Adsorption characteristics of Cd (II) and Cr (VI). *Arab. J. Chem.* **2024**, *17*, 105926. [CrossRef]
80. Frutos, I.; García-Delgado, C.; Gárate, A.; Eymar, E. Biosorption of heavy metals by organic carbon from spent mushroom substrates and their raw materials. *Int. J. Environ. Sci. Technol.* **2016**, *13*, 2713–2720. [CrossRef]
81. Abdallah, M.M.; Ahmad, M.N.; Walker, G.; Leahy, J.J.; Kwapinski, W. Batch and continuous systems for Zn, Cu, and Pb metal ions adsorption on spent mushroom compost biochar. *Ind. Eng. Chem. Res.* **2019**, *58*, 7296–7307. [CrossRef]
82. Kazem, G.G.; Saeedeh, R. Biochar modification and application to improve soil fertility and crop productivity. *Agriculture* **2022**, *68*, 45–61.
83. Leng, L.; Xiong, Q.; Yang, L.; Li, H.; Zhou, Y.; Zhang, W.; Jiang, S.; Li, H.; Huang, H. An overview on engineering the surface area and porosity of biochar. *Sci. Total Environ.* **2021**, *763*, 144204. [CrossRef] [PubMed]
84. Quilliam, R.S.; Glanville, H.C.; Wade, S.C.; Jones, D.L. Life in the 'charosphere'—Does biochar in agricultural soil provide a significant habitat for microorganisms? *Soil Biol. Biochem.* **2013**, *65*, 287–293. [CrossRef]
85. Brewer, C.E.; Brown, R.C. Biochar. In *Comprehensive Renewable Energy*; Elsevier: Amsterdam, The Netherlands, 2012; pp. 357–384.
86. Yang, L.; Li, S.; Ahmed, W.; Jiang, T.; Mei, F.; Hu, X.; Liu, W.; Abbas, F.M.; Xue, R.; Peng, X.; et al. Exploring the relationship between biochar pore structure and microbial community composition in promoting tobacco growth. *Plants* **2024**, *13*, 2952. [CrossRef]
87. Guan, T.K.; Wang, Q.Y.; Li, J.S.; Yan, H.W.; Chen, Q.J.; Sun, J.; Liu, C.J.; Han, Y.N.; Zou, Y.J.; Zhang, G.Q. Biochar immobilized plant growth-promoting rhizobacteria enhanced the physicochemical properties, agronomic characters and microbial communities during lettuce seedling. *Front. Microbiol.* **2023**, *14*, 1218205. [CrossRef] [PubMed]
88. Jatuwong, K.; Aiduang, W.; Kiatsiriroat, T.; Kamopas, W.; Lumyong, S. Effects of biochar and arbuscular mycorrhizal fungi on soil health in Chinese kale (*Brassica oleracea* var. *alboglabra L.)* cultivation. *Microbiol. Res.* **2024**, *15*, 404–421. [CrossRef]
89. Mulyadi; Jiang, L. The combined application of biochar and arbuscular mycorrhizal fungi (AMF) enhanced the physical and chemical properties of soil and rice productivity in Indonesia. *Sustainability* **2023**, *15*, 9782. [CrossRef]
90. Wen, Y.; Wu, R.; Qi, D.; Xu, T.; Chang, W.; Li, K.; Fang, X.; Song, F. The effect of AMF combined with biochar on plant growth and soil quality under saline-alkali stress: Insights from microbial community analysis. *Ecotoxicol. Environ. Saf.* **2024**, *281*, 116592. [CrossRef] [PubMed]
91. Deng, S.; An, Q.; Song, J.; Yang, Y.; Huang, Z.; Feng, S.; Tang, C.; Zhao, B. Enhancement of Mn^{2+}, Fe^{2+} and NH^{4+}-N removal by biochar synergistic strains combined with activated sludge in real wastewater treatment. *Chemosphere* **2024**, *359*, 142271. [CrossRef] [PubMed]
92. Ratheesh, A.; Shibli, S.M.A. Biochar supported *Pseudomonas putida* based globules for effective removal of Bisphenol A with a practical approach. *Chemosphere* **2024**, *361*, 142496. [CrossRef] [PubMed]
93. He, S.; Zhong, L.; Duan, J.; Feng, Y.; Yang, B.; Yang, L. Bioremediation of wastewater by iron oxide-biochar nanocomposites loaded with photosynthetic bacteria. *Front. Microbiol.* **2017**, *8*, 823. [CrossRef]
94. Pandey, D.; Singh, S.V.; Savio, N.; Bhutto, J.K.; Srivastava, R.K.; Yadav, K.K.; Sharma, R.; Nandipamu, T.M.K.; Sarkar, B. Biochar application in constructed wetlands for wastewater treatment: A critical review. *J. Water Process Eng.* **2025**, *69*, 106713. [CrossRef]
95. Osman, A.I.; Fawzy, S.; Farghali, M.; El-Azazy, M.; Elgarahy, A.M.; Fahim, R.A.; Abdel Maksoud, M.I.A.; Ajlan, A.A.; Yousry, M.; Saleem, Y.; et al. Biochar for agronomy, animal farming, anaerobic digestion, composting, water treatment, soil remediation, construction, energy storage, and carbon sequestration: A review. *Environ. Chem. Lett.* **2022**, *20*, 2385–2485. [PubMed]
96. Harris, C. Effect of biochar application on soil microbial communities in degraded lands in Australia. *Int. J. Biol.* **2024**, *4*, 1–12. [CrossRef]
97. Matoso, S.C.G.; Wadt, P.G.S.; Souza, V.S.D.; Pérez, X.L.O. Synthesis of enriched biochar as a vehicle for phosphorus in tropical soils. *Acta Amazon.* **2019**, *49*, 268–276. [CrossRef]

98. Karim, A.A.; Kumar, M.; Singh, E.; Kumar, A.; Kumar, S.; Ray, A.; Dhal, N.K. Enrichment of primary macronutrients in biochar for sustainable agriculture: A review. *Crit. Rev. Environ. Sci. Technol.* **2022**, *52*, 1449–1490. [CrossRef]
99. Bolan, S.; Sharma, S.; Mukherjee, S.; Kumar, M.; Rao, C.S.; Nataraj, K.C.; Singh, G.; Vinu, A.; Bhowmik, A.; Sharma, H.; et al. Biochar modulating soil biological health: A review. *Sci. Total Environ.* **2023**, *914*, 169585. [CrossRef]
100. Kabir, E.; Kim, K.H.; Kwon, E.E. Biochar as a tool for the improvement of soil and environment. *Front. Environ. Sci.* **2023**, *11*, 1324533. [CrossRef]
101. Sharma, P. Biochar application for sustainable soil erosion control: A review of current research and future perspectives. *Front. Environ. Sci.* **2024**, *12*, 1373287. [CrossRef]
102. García-Rodríguez, Á.F.; Moreno-Racero, F.J.; García de Castro Barragán, J.M.; Colmenero-Flores, J.M.; Greggio, N.; Knicker, H.; Rosales, M.A. Influence of biochar mixed into peat substrate on lettuce growth and nutrient supply. *Horticulturae* **2022**, *8*, 1214. [CrossRef]
103. Tay, C.C.; Shaari, M.S.; Anuar, W.N.H.W.; Hashim, S.N. *Pleurotus* spent mushroom compost as green supplementary nutrient in tissue culture. *MATEC Web Conf.* **2016**, *47*, 05010. [CrossRef]
104. Wu, Q.; Xian, Y.; He, Z.; Zhang, Q.; Wu, J.; Yang, G.; Zhang, X.; Qi, H.; Ma, j.; Xioa, Y.; et al. Adsorption characteristics of Pb (II) using biochar derived from spent mushroom substrate. *Sci. Rep.* **2019**, *9*, 15999. [CrossRef] [PubMed]
105. Zhao, W.; Wang, W.; Wu, Y.; Guo, W.; Ren, W.; Meng, S.; Yun, S.; Feng, C. Magnetic biochar prepared from a spent mushroom substrate as an adsorbent for the analysis of pyrethroids in environmental water samples. *Talanta* **2024**, *284*, 127195. [CrossRef] [PubMed]
106. Elad, Y.; David, D.R.; Harel, Y.M.; Borenshtein, M.; Kalifa, H.B.; Silber, A.; Graber, E.R. Induction of systemic resistance in plants by biochar, a soil-applied carbon sequestering agent. *Phytopathology* **2010**, *100*, 913–921. [CrossRef] [PubMed]
107. Frenkel, O.; Jaiswal, A.K.; Elad, Y.; Lew, B.; Kammann, C.; Graber, E.R. The effect of biochar on plant diseases: What should we learn while designing biochar substrates? *J. Environ. Eng. Landsc. Manag.* **2017**, *25*, 105–113. [CrossRef]
108. Babu, S.; Singh, R.; Kumar, S.; Rathore, S.S.; Yadav, D.; Yadav, S.K.; Yadav, V.; Ansari, M.A.; Das, A.; Rajanna, G.A.; et al. Biochar implications in cleaner agricultural production and environmental sustainability. *Environ. Sci. Adv.* **2023**, *2*, 1042–1059. [CrossRef]
109. Xiang, L.; Harindintwali, J.D.; Wang, F.; Redmile-Gordon, M.; Chang, S.X.; Fu, Y.; He, C.; Muhoza, B.; Brahushi, F.; Bolan, N.; et al. Integrating biochar, bacteria, and plants for sustainable remediation of soils contaminated with organic pollutants. *Environ. Sci. Technol.* **2022**, *56*, 16546–16566. [CrossRef] [PubMed]
110. Ajema, L. Effects of biochar application on beneficial soil organism. *Int. J. Res. Stud. Sci. Eng. Technol.* **2018**, *5*, 9–18.
111. Ahmad, M.; Ok, Y.S.; Kim, B.Y.; Ahn, J.H.; Lee, Y.H.; Zhang, M.; Moon, D.H.; Al-Wabel, M.I.; Lee, S.S. Impact of soybean stover-and pine needle-derived biochars on Pb and As mobility, microbial community, and carbon stability in a contaminated agricultural soil. *J. Environ. Manag.* **2016**, *166*, 131–139. [CrossRef] [PubMed]
112. Jin, X.; Zhou, X.; Wu, F.; Xiang, W.; Pan, K. Biochar amendment suppressed Fusarium wilt and altered the rhizosphere microbial composition of tomatoes. *Agronomy* **2023**, *13*, 1811. [CrossRef]
113. Murtaza, G.; Ahmed, Z.; Eldin, S.M.; Ali, B.; Bawazeer, S.; Usman, M.; Iqbal, R.; Neupane, D.; Ullah, A.; Khan, A.; et al. Biochar-Soil-Plant interactions: A cross talk for sustainable agriculture under changing climate. *Front. Environ. Sci.* **2023**, *11*, 1059449. [CrossRef]
114. Bhattarai, R.; Karki, N.; Shakya, S.; Dhakal, R.P.; Poudel, P. Potential application of biochar as a growth supplement for mushroom cultivation (*Pleurotus ostreatus*). *Int. J. Hortic. Food Sci.* **2024**, *6*, 21–26. [CrossRef]
115. Zakil, F.A.; Sueb, M.S.M.; Isha, R.; Kamaluddin, S.H. Efficiency of charcoal as supporting growth material in *Pleurotus ostreatus* mushroom cultivation on various agricultural wastes mixed with rubber tree sawdust (SR). *Chem. Eng. Trans.* **2021**, *89*, 415–420.
116. Mahari, W.A.W.; Peng, W.; Nam, W.L.; Yang, H.; Lee, X.Y.; Lee, Y.K.; Liew, R.K.; Ma, N.L.; Mohammad, A.; Sonne, C.; et al. A review on valorization of oyster mushroom and waste generated in the mushroom cultivation industry. *J. Hazard. Mater.* **2020**, *400*, 123156. [CrossRef]
117. FML Engineering. Biochar Boosting Mushroom Cultivation. Available online: https://www.fmlengineering.com/renewable-energy/biochar/biochar-boosting-mushroom-cultivation/ (accessed on 18 November 2024).
118. Kansara, H.; Manipati, M.; Hernandez-Charpak, Y.; Carter, C.; Luft, L.; Hallowell, J.; Trabold, T. Achieving Circularity in the Mushroom Industry: Biochar from Spent Mushroom Substrate. Available online: https://biochar-us.org/usbi-nabc24-circularitywithbiocharproductionfromspentmushroomsubstrates (accessed on 18 November 2024).
119. Tomczyk, A.; Sokołowska, Z.; Boguta, P. Biochar physicochemical properties: Pyrolysis temperature and feedstock kind effects. *Rev. Environ. Sci. Biotechnol.* **2020**, *19*, 191–215. [CrossRef]
120. Ravindiran, G.; Rajamanickam, S.; Janardhan, G.; Hayder, G.; Alagumalai, A.; Mahian, O.; Lam, S.S.; Sonne, C. Production and modifications of biochar to engineered materials and its application for environmental sustainability: A review. *Biochar* **2024**, *6*, 62. [CrossRef]
121. Wijitkosum, S. Biochar derived from agricultural wastes and wood residues for sustainable agricultural and environmental applications. *Int. Soil. Water. Conserv. Res.* **2022**, *10*, 335–341. [CrossRef]

122. Ayaz, M.; Feizienė, D.; Tilvikienė, V.; Akhtar, K.; Stulpinaitė, U.; Iqbal, R. Biochar role in the sustainability of agriculture and environment. *Sustainability* **2021**, *13*, 1330. [CrossRef]
123. Nan, H.; Yin, J.; Yang, F.; Luo, Y.; Zhao, L.; Cao, X. Pyrolysis temperature-dependent carbon retention and stability of biochar with participation of calcium: Implications to carbon sequestration. *Environ. Pollut.* **2021**, *287*, 117566. [CrossRef] [PubMed]
124. Yin, X.; Peñuelas, J.; Xu, X.; Sardans, J.; Fang, Y.; Wiesmeier, M.; Chen, Y.; Chen, X.; Wang, W. Effects of addition of nitrogen-enriched biochar on bacteria and fungi community structure and C, N, P, and Fe stoichiometry in subtropical paddy soils. *Eur. J. Soil. Biol.* **2021**, *106*, 103351. [CrossRef]
125. Kasera, N.; Kolar, P.; Hall, S.G. Nitrogen-doped biochars as adsorbents for mitigation of heavy metals and organics from water: A review. *Biochar* **2022**, *4*, 17. [CrossRef]
126. Grimm, A.; dos Reis, G.S.; Khokarale, S.G.; Ekman, S.; Lima, E.C.; Xiong, S.; Hultberg, M. Shiitake spent mushroom substrate as a sustainable feedstock for developing highly efficient nitrogen-doped biochars for treatment of dye-contaminated water. *J. Water Process Eng.* **2023**, *56*, 104435. [CrossRef]
127. Yaashikaa, P.R.; Kumar, P.S.; Varjani, S.; Saravanan, A.J.B.R. A critical review on the biochar production techniques, characterization, stability and applications for circular bioeconomy. *Biotechnol. Rep.* **2020**, *28*, e00570. [CrossRef]
128. Alhashimi, H.A.; Aktas, C.B. Life cycle environmental and economic performance of biochar compared with activated carbon: A meta-analysis. *Resour. Conserv. Recycl.* **2017**, *118*, 13–26. [CrossRef]
129. Zhang, P.; Duan, W.; Peng, H.; Pan, B.; Xing, B. Functional biochar and its balanced design. *ACS Environ. Au* **2021**, *2*, 115–127. [CrossRef] [PubMed]
130. Zhang, T.; Han, J.D.; Li, J. Comprehensive utilization and research status of edible mushrooms residue. *Shandong Agric. Sci.* **2016**, *48*, 146–150.
131. Nematian, M.; Keske, C.; Ng'ombe, J.N. A techno-economic analysis of biochar production and the bioeconomy for orchard biomass. *Waste Manag.* **2021**, *135*, 467–477. [CrossRef]
132. Alsawy, T.; Rashad, E.; El-Qelish, M.; Mohammed, R.H. A comprehensive review on the chemical regeneration of biochar adsorbent for sustainable wastewater treatment. *NPJ Clean. Water* **2022**, *5*, 29. [CrossRef]
133. Phiri, Z.; Moja, N.T.; Nkambule, T.T.; de Kock, L.A. Utilization of biochar for remediation of heavy metals in aqueous environments: A review and bibliometric analysis. *Heliyon* **2024**, *10*, e25785. [CrossRef] [PubMed]
134. Sabio, E.; González, E.; González, J.F.; González-Garcıía, C.M.; Ramiro, A.; Ganan, J. Thermal regeneration of activated carbon saturated with p-nitrophenol. *Carbon* **2004**, *42*, 2285–2293. [CrossRef]
135. San Miguel, G.; Lambert, S.D.; Graham, N.J.D. The regeneration of field-spent granular-activated carbons. *Water Res.* **2001**, *35*, 2740–2748. [CrossRef] [PubMed]
136. Liao, W.; Drake, J.; Thomas, S.C. Biochar granulation enhances plant performance on a green roof substrate. *Sci. Total Environ.* **2022**, *813*, 152638. [CrossRef]
137. Barbhuiya, S.; Das, B.B.; Kanavaris, F. Biochar-concrete: A comprehensive review of properties, production and sustainability. *Case Stud. Constr. Mater.* **2024**, *20*, e02859. [CrossRef]
138. Senadheera, S.S.; Withana, P.A.; Lim, J.Y.; You, S.; Chang, S.X.; Wang, F.; Rhee, J.H.; Ok, Y.S. Carbon negative biochar systems contribute to sustainable urban green infrastructure: A critical review. *Green Chem.* **2024**, *26*, 10634–10660. [CrossRef]
139. Kader, S.; Gratchev, I.; Michael, R.N. Recycled waste substrates: A systematic review. *Sci. Total Environ.* **2024**, *953*, 176029. [CrossRef] [PubMed]
140. Cao, C.T.; Farrell, C.; Kristiansen, P.E.; Rayner, J.P. Biochar makes green roof substrates lighter and improves water supply to plants. *Ecol. Eng.* **2014**, *71*, 368–374. [CrossRef]
141. Werdin, J.; Conn, R.; Fletcher, T.D.; Rayner, J.P.; Williams, N.S.; Farrell, C. Biochar particle size and amendment rate are more important for water retention and weight of green roof substrates than differences in feedstock type. *Ecol. Eng.* **2021**, *171*, 106391. [CrossRef]
142. Chen, H.; Ma, J.; Wei, J.; Gong, X.; Yu, X.; Guo, H.; Zhao, Y. Biochar increases plant growth and alters microbial communities via regulating the moisture and temperature of green roof substrates. *Sci. Total Environ.* **2018**, *635*, 333–342. [CrossRef] [PubMed]
143. Qianqian, Z.; Liping, M.; Huiwei, W.; Long, W. Analysis of the effect of green roof substrate amended with biochar on water quality and quantity of rainfall runoff. *Environ. Monit. Assess.* **2019**, *191*, 304. [CrossRef] [PubMed]
144. Lee, J.; Kwon, E.E. Biochar in Green Roofs: A Review. *J. Build. Eng.* **2024**, *89*, 109272. [CrossRef]
145. Mustafa, A.; Brtnicky, M.; Hammerschmiedt, T.; Kucerik, J.; Kintl, A.; Chorazy, T.; Naveed, M.; Skarpa, P.; Baltazar, T.; Malicek, O.; et al. Food and agricultural wastes-derived biochars in combination with mineral fertilizer as sustainable soil amendments to enhance soil microbiological activity, nutrient cycling and crop production. *Front. Plant Sci.* **2022**, *13*, 1028101. [CrossRef] [PubMed]
146. Yadav, S.P.S.; Bhandari, S.; Bhatta, D.; Poudel, A.; Bhattarai, S.; Yadav, P.; Ghimire, N.; Paudel, P.; Paudel, P.; Shrestha, J.; et al. Biochar application: A sustainable approach to improve soil health. *J. Agric. Food Res.* **2023**, *11*, 100498.

147. Patel, R.; Babaei-Ghazvini, A.; Dunlop, M.J.; Acharya, B. Biomaterials-based concrete composites: A review on biochar, cellulose and lignin. *Carbon Capture Sci. Technol.* **2024**, *12*, 100232. [CrossRef]
148. Osman, A.I.; Farghali, M.; Dong, Y.; Kong, J.; Yousry, M.; Rashwan, A.K.; Chen, Z.; Al-Fatesh, A.; Rooney, D.W.; Yap, P.S. Reducing the carbon footprint of buildings using biochar-based bricks and insulating materials: A review. *Environ. Chem. Lett.* **2024**, *22*, 71–104. [CrossRef]
149. Nair, P.S.; Sivani, M.P.; Suresh, S.; Sreekanth, A.J.; Sivasabari, K.; Adithya, K.S.; Anuranj, P.R.; Krishnan, N.; Parvathy, S.; Chakraborty, S.; et al. Beneficial impacts of biochar as a potential feed additive in animal husbandry. *J. Exp. Biol. Agric.* **2023**, *11*, 479–499. [CrossRef]
150. Man, K.Y.; Chow, K.L.; Man, Y.B.; Mo, W.Y.; Wong, M.H. Use of biochar as feed supplements for animal farming. *Crit. Rev. Environ. Sci. Technol.* **2021**, *51*, 187–217. [CrossRef]
151. Graves, C.; Kolar, P.; Shah, S.; Grimes, J.; Sharara, M. Can biochar improve the sustainability of animal production? *Appl. Sci.* **2022**, *12*, 5042. [CrossRef]
152. Mainali, K.; Mood, S.H.; Pelaez-Samaniego, M.R.; Sierra-Jimenez, V.; Garcia-Perez, M. Production and applications of N-doped carbons from bioresources: A review. *Catal. Today* **2023**, *423*, 114248. [CrossRef]
153. Saravanan, A.; Maruthapandi, M.; Das, P.; Ganguly, S.; Margel, S.; Luong, J.H.; Gedanken, A. Applications of N-doped carbon dots as antimicrobial agents, antibiotic carriers, and selective fluorescent probes for nitro explosives. *ACS Appl. Bio Mater.* **2020**, *3*, 8023–8031. [CrossRef] [PubMed]
154. Ansari, L.; Hallaj, S.; Hallaj, T.; Amjadi, M. Doped-carbon dots: Recent advances in their biosensing, bioimaging and therapy applications. *Colloids Surf. B Biointerfaces* **2021**, *203*, 111743. [CrossRef] [PubMed]
155. Liu, Y.; Ma, R.; Li, D.; Qi, C.; Han, L.; Chen, M.; Fu, F.; Yuan, J.; Li, G. Effects of calcium magnesium phosphate fertilizer, biochar and spent mushroom substrate on compost maturity and gaseous emissions during pig manure composting. *J. Environ. Manag.* **2020**, *267*, 110649. [CrossRef]
156. Horn, L.; Ueitele, I.S.; Nghifewa, M. Exploring various promising green strategy for recycling spend mushroom substrate through desktop research analysis. *Int. J. Environ. Sci. Nat. Resour.* **2021**, *29*, 556251.
157. Da Silva Medeiros, D.C.C.; Nzediegwu, C.; Benally, C.; Messele, S.A.; Kwak, J.H.; Naeth, M.A.; Ok, Y.S.; Chang, S.X.; El-Din, M.G. Pristine and engineered biochar for the removal of contaminants co-existing in several types of industrial wastewaters: A critical review. *Sci. Total Environ.* **2022**, *809*, 151120. [CrossRef]

MDPI AG
Grosspeteranlage 5
4052 Basel
Switzerland
Tel.: +41 61 683 77 34

Life Editorial Office
E-mail: life@mdpi.com
www.mdpi.com/journal/life